HeadStart Primary

Maths Problem Solving, Reasoning & Investigating

Year 3

Lizzie Marsland
Susannah Palmer

Acknowledgements:

Author: Lizzie Marsland, Susannah Palmer

Series Editor: Peter Sumner

Cover and Page Design: Kathryn Webster, Jo Sullivan

The right of Lizzie Marsland and Susannah Palmer to be identified as the authors of this publication has been asserted by them in accordance with the Copyright, Designs and Patents Act 1998.

HeadStart Primary Ltd
Elker Lane
Clitheroe
BB7 9HZ

T. 01200 423405
E. info@headstartprimary.com
www.headstartprimary.com

Published by HeadStart Primary Ltd 2018

A record for this book is available from the British Library -
ISBN: 978-1-908767-59-2

CONTENTS

CONTENTS

Year 3

INTRODUCTION

These problems have been written in line with the objectives from the Mathematics Curriculum. Questions have been written to match all appropriate objectives from each content domain of the curriculum.

Solving problems and mathematical reasoning in context are difficult skills for children to master; a real-life, written problem is an abstract concept and children need opportunities to practise and consolidate their problem solving techniques.

As each content domain is taught, the skills learnt can be applied to the relevant problems. This means that a particular objective can be reinforced and problem solving and reasoning skills further developed. The first section of each content domain is intended to provide opportunities for children to practise and consolidate their problem solving skills. Each page has an identified objective from the National Curriculum; the difficulty level of the questions increases towards the bottom of each page, thus providing built-in differentiation.

Mastering a skill involves obtaining a greater level of understanding of the skill, the ability to transfer and apply knowledge in different contexts and explaining understanding to others.

The MASTERING and INVESTIGATION sections provide extra challenges as children's problem solving skills and confidence increase. The problems in the MASTERING sections encompass several objectives from the relevant curriculum domain. The INVESTIGATION covers objectives from across the whole curriculum.

At HeadStart, we realise that children may need more space to record their answers, working out or explanations. It is recommended, therefore, that teachers use their discretion as to where children complete their work.

Since a structured approach to problem solving supports learning, developing a whole-school approach is highly recommended.

Throughout this book, 6 children are solving problems.

Their names are:

Sarah

Kyle

Jessica

Siddiq

Humma

Isiah

It may be appropriate for children to use exercise books or paper to record their answers, working out or explanations.

NUMBER

Number and place value

Count from 0 in multiples of 4, 8, 50 and 100

1 Kyle is counting in multiples of **4**. The **first** number he says is **0**. What is the **third** number he says?

2 Sarah is making an apple pie. She already has **4** apples and then buys **4** more. Her grandma also gives her **4** apples. How many apples does she have altogether?

3 Jessica counts **eight** birds in the sky. She then sees **8** more. How many birds has she counted altogether?

4 Mrs Brown asked her class to count in multiples of **8**. The **first** number they said was **8**. What was the **fourth** number they said?

5 Humma and Isiah are counting in multiples of **50**. They start counting at **0**. What are the next **two** numbers they count?

6 Jessica and Siddiq counted the cows in **3** fields. Each field had **100** cows eating grass. How many cows were there altogether?

Name ....

Count from 0 in multiples of 4, 8, 50 and 100

1 Sarah is counting in multiples of **four**. The **first** number she says is **4**. What is the **third** number she says?

☐

2 Humma has **8** colouring pencils. Jessica has **sixteen**. How many more colouring pencils does Jessica have?

☐

3 Kyle and Humma like to count in **hundreds** together. If the **first** number they said was **0**, what was the **fourth** number they said?

☐

4 Jessica is counting in multiples of **50**. She starts at **0**. Write the next **three** numbers that she counts.

☐ ☐ ☐

5 Isiah and **four** of his friends are taking turns to count up in steps of **100**. If Isiah starts counting at **0**, what number would they count up to?

☐

6 Mrs Taylor writes this number pattern on the whiteboard. Can you fill in the missing numbers?

0 **4** **8** ☐ **16** ☐ ☐

Name ....

Count in and use multiples of 2, 3, 4, 5, 50 and 100

1. Siddiq is counting in multiples of **two**. His **first** number is **2**. What would be the next number he counts?

2. How many multiples of **four** should Sarah be able to find that are less than **19**?

3. Jessica has **4** cakes. Isiah has **4** cakes and Humma has **4** cakes. How many cakes do they have altogether?

4. Mrs Whelan asks her class how many multiples of **five** they can count between **50** and **100**. What did they tell her?

5. Jessica is counting in multiples of **100**. Her **first three** numbers are **566**, **666** and **766**. What would the next number in her sequence be?

6. Kyle thinks of a number and multiplies it by **3**. His answer is **15**. What was the number Kyle thought of?

Name ....

Count in and use multiples of 2, 3, 4, 5, 50 and 100

1. Isiah counts in multiples of **three**. He starts at number **6**. What number does he count next?

2. Year 3 are counting in multiples of **two**. They start at number **10**. What are the next **three** numbers they say?

3. Sarah has **25** teddy bears. She gets **5** more teddy bears for her birthday. How many teddy bears does she have now?

4. Siddiq is counting up in **fours**. He starts at number **16**. Which **two** numbers does he say next?

5. Kyle starts at the number **40** and he counts up **three** lots of **fifty**. Which number does he stop at?

6. Miss Price asks her class to count in multiples of **100**, starting from **132**. Which **four** numbers should they say next?

Name ....

Find 10 or 100 more or less than a given number

1. Humma has **200** raisins and eats **100** of the raisins at lunchtime, how many does she have left for home time?

2. Siddiq has **87** marbles. He gives **10** to his friend, Isiah. How many marbles does Siddiq have left?

3. Sarah counts forward **ten**. She starts at **64**. What number does she count up to?

4. There are **25** candles on the birthday cake. Kyle blows out **10** of the candles. How many burning candles are left?

5. Isiah has collected **224** football cards. For his birthday, his dad buys him another **100** cards. How many does he have now?

6. Sarah is **seven** years old. Her sister is **10** years older than her. How old is her sister?

Name ....

Find 10 or 100 more or less than a given number

1 Jessica has baked **58** cupcakes. Her friends eat **10**. How many cupcakes are left?

2 Mr Peabody asks his class to count **100** less than **347**. What number should the class say?

3 Kyle counts **86** sheep in Field **A**. **10** sheep are moved to Field **B**. How many sheep are left in Field **A**?

4 There are **72** children on the playground. **10** more children go out to play. How many children are on the playground now?

5 Sarah starts at **334** and then counts forward **one hundred**. What number does she count to?

6 Humma has saved **£256**. She spends **£100** on new clothes. How much money does she have left?

£

Name ....

Recognise the place value of each digit in a three-digit number (hundreds, tens and ones)

1 There are **252** pupils at Apple Tree Primary School. How many **hundreds** are there?

---

2 Jessica counts **335** flowers in the park. She then partitions the number of flowers into **hundreds**, **tens** and **ones**. How many **tens** are there?

---

3 Siddiq partitions the number **528** into **hundreds**, **tens** and **ones**. How many **hundreds** are there?

---

4 Miss Bell asks her class to partition **764** into **hundreds**, **tens** and **ones**. How many **ones** are there?

---

5 Humma has to write down the value of the digit **6** in the number **463**. What should she write?

---

6 Isiah has partitioned a number into **hundreds**, **tens** and **ones** and it looks like this:

300 + 20 + 9

What was the number?

Name ....

Recognise the place value of each digit in a three-digit number (hundreds, tens and ones)

1 Jessica partitions **482** into **hundreds**, **tens** and **ones**. How would she write this down?

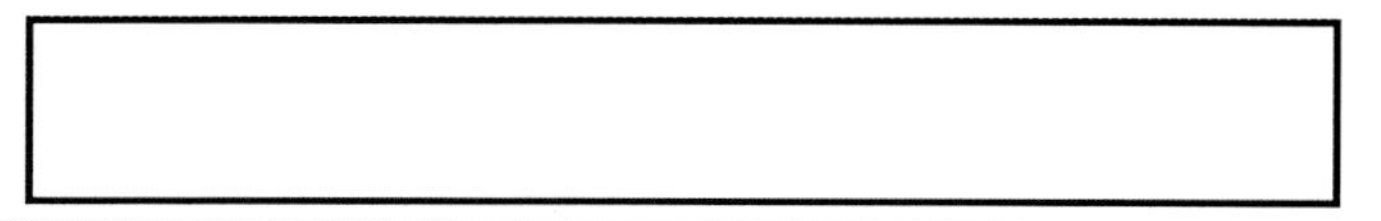

2 Siddiq partitioned the number below. What number did he start with?

200 + 80 + 7

3 Siddiq partitions another number in a different way. What was the number?

700 + 30 + 16

4 Sarah's dad has forgotten about place value. He tells Sarah that the digit **4** in the number **649** is worth **400**. Is he correct? Explain your answer.

Yes / No

....

....

5 Isiah says, "The value of the digit **5** in the number **562** is **50**." Is he correct? Explain your answer.

Yes / No

....

....

Name ....

Compare and order numbers up to 1000

1. Humma had to find out which number was smaller: **482** or **428**. Which one did she choose? How did she know?

   ....

   ....

2. Jessica put the following numbers in order from largest to smallest. What does her new list look like?

310	305	300	320
largest			smallest

3. Kyle's teacher asks him to write a **three-digit** number that is larger than **535**, but smaller than **537**. What number should he write?

4. Write down a number that is smaller than **350**, but larger than **348**.

5. Miss Collins asks her class to put the following numbers in order of size, starting with the smallest. Can you help?

345	435	354	543	534
smallest				largest

Name ....

Compare and order numbers up to 1000

1 Here is a sequence of numbers. Put the numbers in order of size from smallest to largest.

882 **820** **865** **840**

smallest **largest**

---

2 Jessica is trying to decide which number is larger: **895** or **859**. Which one should she choose? How would she know?

....

....

---

3 Kyle puts these numbers in order of size from largest to smallest. What do you think his new list looks like?

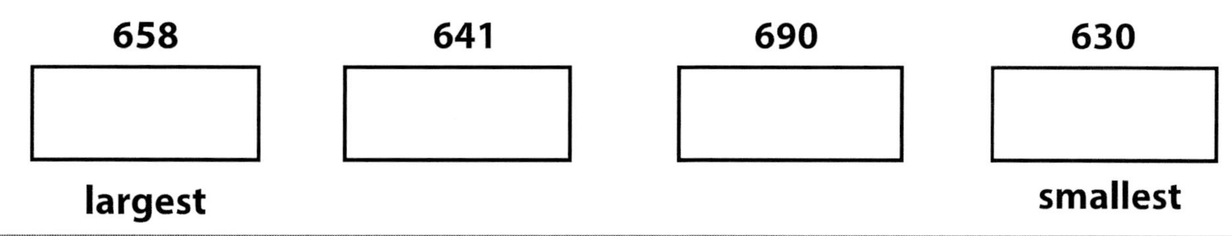

---

4 Mr Watson asked his class to write down a number which was larger than **489** but smaller than **491**. What number should they choose?

---

5 Sarah wants to put these numbers in order of size from largest to smallest. What does her new list of numbers look like?

981 **918** **891** **819** **899**

largest **smallest**

Name ....

Identify, represent and estimate numbers using different representations

1 Siddiq counts **26** bees making honey. Is this number of bees nearer to **20** or **30**?

2 Isiah jumps **6 metres** in the long jump. Humma jumps **7 metres**. Who jumps further?

3 Kyle throws a ball **13 metres** for his dog, Bruno. Is this nearer to **10 metres** or **20 metres**?

4 There are **70** children singing in the school choir. **36** are girls and **34** are boys. Are there more girls or boys singing in the choir?

5 Mr Rajan asks Isiah what number comes exactly halfway between **40** and **60**. What number should Isiah say?

6 Jessica is **115 cm** tall. Her friend, Sarah, is **10 cm** taller than her. Humma is **135 cm** tall. Who is the tallest?

Name ....

Identify, represent and estimate numbers using different representations

1 Mrs Smith gave **145 millilitres** of milk to her cat. Was this amount nearer to **100** or **200 millilitres**?

[ ] **ml**

2 Siddiq says, "The number which comes exactly halfway between **60** and **80** is **70**." Is Siddiq correct? Explain your answer.

....

....

3 There are **796** fans cheering on Redton United. Is the number of fans nearer to **790** or **800**? Explain your answer.

....

....

4 **26** boys and **24** girls went to the school party. How many boxes of blackcurrant juice cartons should Mr Foster buy to give the children a drink each, if blackcurrant juice cartons come in boxes of **10**?

[ ]

5 Isiah has a bottle containing **786 millilitres** of orange juice. He drinks **100 millilitres**. How much orange juice does he have left in the bottle, to the nearest **100 ml**?

[ ] **ml**

Name ....

Read and write numbers up to 1000 in numerals and words

1. Mr Baker asks his class to write the number **72** in words. How would the class write this number?

   ....

2. Sarah's teacher asks her to write the number **two hundred and thirty** in numerals. What should Sarah write?

3. Humma lives at number **three hundred and forty three**. Her dad is making a sign for their home. Which numerals should he write on the sign?

4. Kyle writes the number **eight hundred and sixty two** in numerals. Which numerals does he write?

5. Isiah writes the number **787** in words. What does he write?

   ....

6. Sarah's mum asks her to write the number **853** in words. What should she write?

   ....

Name ....

Read and write numbers up to 1000 in numerals and words

1. Kyle is writing a birthday card for his friend who is going to be **8**. How would he write this number in the card using a word?

   ....

2. Jessica counts **fifteen** birds in the sky. How would she write this as a numeral?

3. Mr Iqbal asked Humma to write the number **252** on the whiteboard in words. What should she write?

   ....

4. Siddiq reads the number **four hundred and sixty four**. How would he write this using numerals?

5. Kyle writes the number **348** in words. Which words does he write?

   ....

6. Jessica's little sister has forgotten how to write the number **876** in words. What are the words that she should write?

   ....

Name ....

Solve problems involving number and place value (money)

1 How much change does Isiah get from **£1**, if he buys a lollipop for **10p**?

p

2 Sarah has **267 pence** in her money jar. Write this as an amount of pounds.

£

3 Mum has **£300**. She gives **£100** to each of her daughters, Humma and Sarah. How much money does Mum have left?

£

4 Which amount of money is bigger: **£389** or **£398**?

£

5 If you could choose, would you rather have the value of the **5** in **£650** or the value of the **5** in **£975**?

6 Mrs Jenkins sees that a washing machine is on offer for **one hundred and twenty six pounds**. How would you write this using numbers and the pound sign?

£

Name ....

Solve problems involving number and place value (distance and capacity)

1 A bottle holds **100 millilitres** of water. How much water is left if Siddiq drinks **10 millilitres**?

[ ] **ml**

2 Humma swam **6** lengths on Monday, **2** more than this on Wednesday and **7** lengths on Thursday. On which day did she swim the most lengths?

[ ]

3 Sarah has a piece of string which is **123 cm** long. Isiah has a piece of string which is **132 cm** long. Who has the longer piece of string?

[ ]

4 Kyle is going on holiday. His suitcase weighs **14 kg**. Is the weight of his suitcase nearer to **10 kg** or **20 kg**?

[ ] **kg**

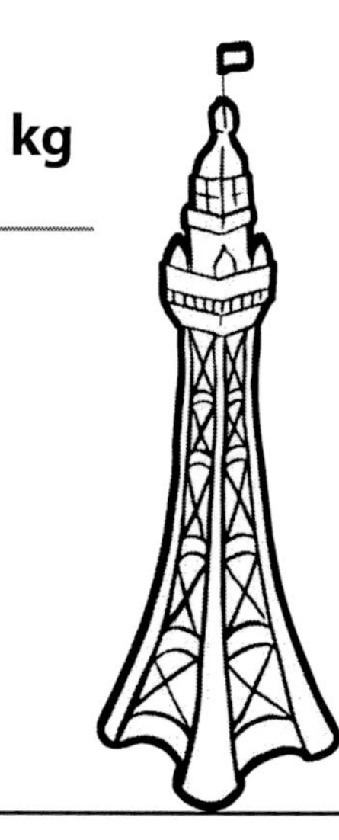

5 Isiah's dad drives **267** miles to Blackpool. On the way back, he goes a different way and drives **276** miles. Which route was shorter? Circle your answer.

267 miles or **276 miles**

Name ....

Solve mixed problems involving number and place value

1. There are **120** crisps in a big bag. Kyle shares them with his friends and they eat **100** crisps altogether. How many crisps are left?

2. **26** children go on a camping trip. Is the number of children on the trip nearer to **20** or **30**?

3. There are **36** bouncy balls in one tub and **10** in another tub. How many are there altogether?

4. Isiah wrote down a number that is more than **200** and a multiple of **four**. It is less than **205**. What number did he write?

5. Joe is **18** and Henry is **12** years younger than Joe. Their cousin Caitlin is **14**. Who is the eldest and who is the youngest?

eldest =   **youngest =**

6. Sarah has a good way of checking whether a whole number is a multiple of **5**. What do you think her method is?

....

....

Name ....

Solve mixed problems involving number and place value

1. Siddiq is counting forward in **tens**. He starts at **92** and then counts on **2** more **tens**. What number does he count up to?

2. Write the number **256** in words.

   ....

3. Siddiq puts **500 grams** of soil in his plant pot. Then he adds another **4** scoops of soil, each containing **100 grams**. How many **grams** of soil are in the plant pot now?

   g

4. Sarah has partitioned a number into **hundreds**, **tens** and **ones** and it looks like this: **600 + 40 + 3**.

   What was the number?

5. Humma has written **36** party invitations. She hands out **10** invitations out to her friends at school and **10** to her friends at Brownies. How many invitations does she have left to give out?

Name ....

Solve mixed problems involving number and place value

1. Sarah counts **46** cows in a field. Is the number of cows nearer to **40** or **50**?

2. Mr Johnson asks Isiah to write the number **four hundred and fifty two** in numerals. What does he write?

3. Jessica's teacher asks her to partition **929** into **hundreds**, **tens** and **ones**. How many **hundreds** are there?

4. If Isiah starts at the number **94** and keeps subtracting **10**, what is the smallest number he would reach on a **100** square?

5. Humma partitions the number **128**. Show **two** ways that she could write this.

6. Sarah thinks of a secret number. It is a multiple of **8**. The number is larger than **30** but smaller than **35**. What is Sarah's number?

Name ....

Solve mixed problems involving number and place value

1. Isiah has **34** stickers. He gives **10** to his friend, Siddiq. How many stickers does Isiah have left?

2. Siddiq writes the number **143** in words. How does he write this?

   ....

3. There are **226** children eating their lunch. **100** children are having a packed lunch and the rest are having a school lunch. How many children are having a school lunch?

4. Sarah is making pancakes. She puts **200 g** of flour into a bowl. She then adds **6** more tablespoons of flour, each containing **10 g** of flour. How much flour is in the bowl now?

   g

5. Kyle partitions **985** in **two** different ways. How might he do this?

6. Look at the numbers below. What is the difference between the smallest and the largest number?

   760 **710** **780** **720** **790**

It may be appropriate for children to use exercise books or paper to record their answers, working out or explanations.

MASTERING

Number and place value

1 Look at the abacus pictures.

a Write the values in the boxes underneath.

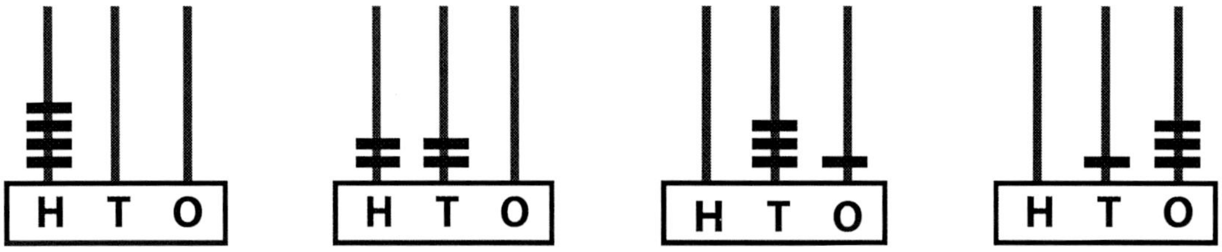

H T O | H T O | H T O | H T O

b All the numbers above have been made by using **four** hoops on the abacus. How many more numbers can you make using **four** hoops?

Can you check that you have found them all?

c Which is the largest number you have made?
How do you know this is the largest possible number?

....

d Which is the smallest number you have made?
How do you know this is the smallest possible number?

....

Name ....

2 What is wrong with this sequence of numbers?

123 **133** **143** **163** **173**

....

....

3 Isiah, Kyle, Siddiq and Sarah each have a raffle ticket.

757	557	75	767

Isiah says: My number has **5** hundreds.

Kyle says: My number is a **2-digit** number.

Siddiq says: My number is the largest number.

Sarah says: The **tens** digit is **2** smaller than the **hundreds** digit.

Who has which number?

Isiah [ ] **Siddiq** [ ]

Kyle [ ] **Sarah** [ ]

Name ....

4 Use lines to join the boxes that match.

All multiples of 50 are multiples of 100.	**true**
All multiples of 100 are multiples of 50.	**false**
All multiples of 50 are odd.	**false**

5 Mrs Smith is making a new display for the maths board in the Year 3 classroom. Every **50 cm** she puts a **2D** shape around the board.

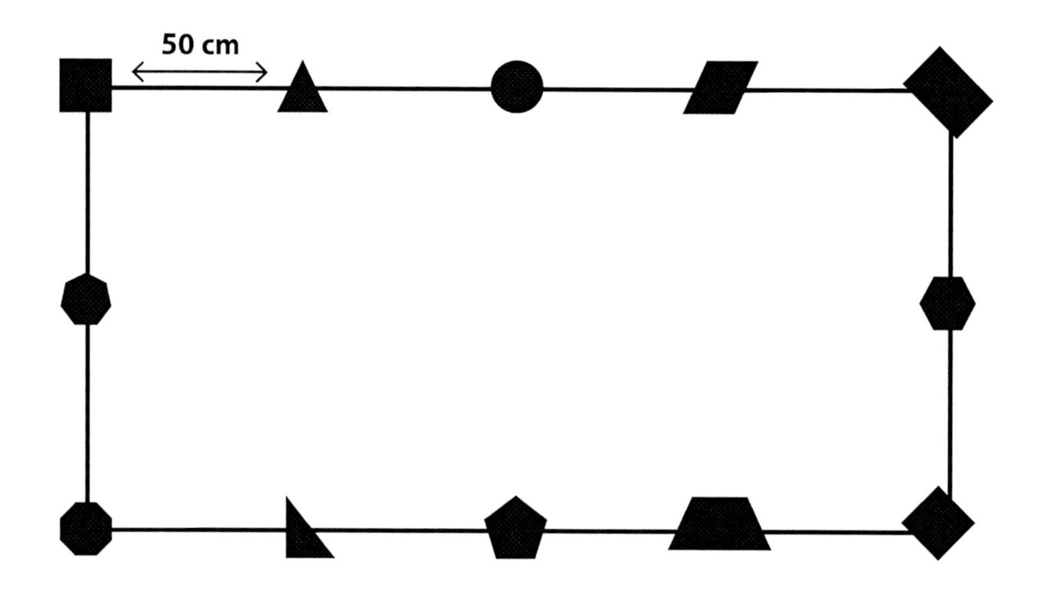

a How long is the board?

☐ **cm**

b What length of ribbon is needed to go around the whole board?

☐ **cm**

Name ....

6

365 = 3 hundreds + 6 tens + 5 ones
365 = 36 tens + 5 ones
365 = 3 hundreds + 65 ones

Show Jessica how to partition 837 in three different ways.

837 = [ ]

837 = [ ]

837 = [ ]

7

a Write the numbers below in the correct place in the Venn diagram.

10 15 16 20 24 25 32 40

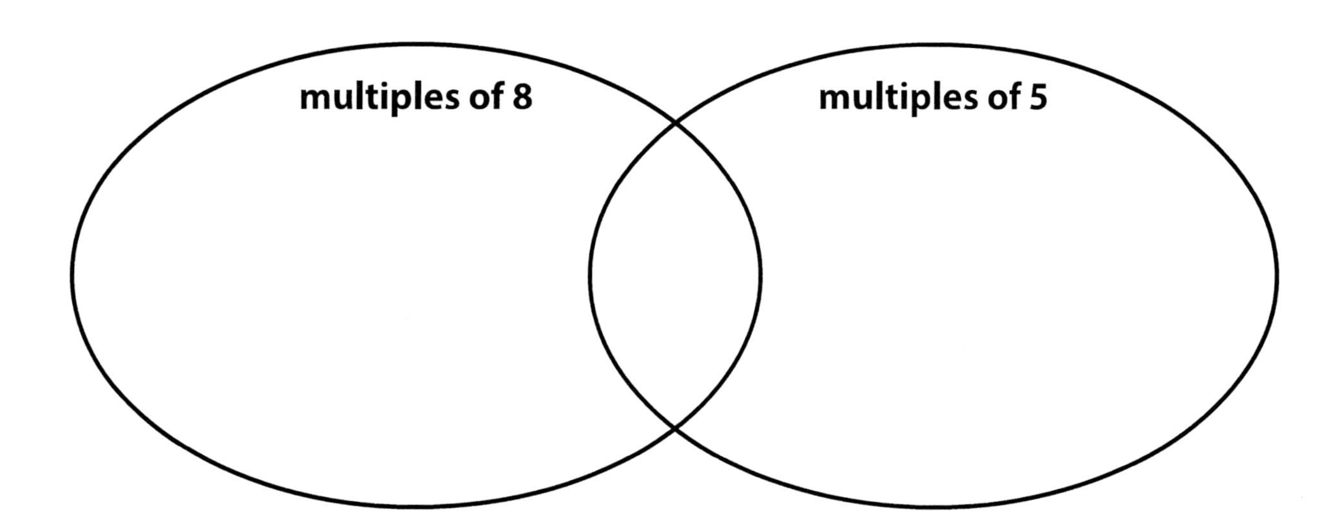

b Where would you put the number **23**?

....

c Write another number in the section where you have put **40**.

Name ....

8 Sarah arranges some number cards to make **3-digit** numbers.

Use the same digit in each row to make the statement correct.

Can you do this **3** different ways?

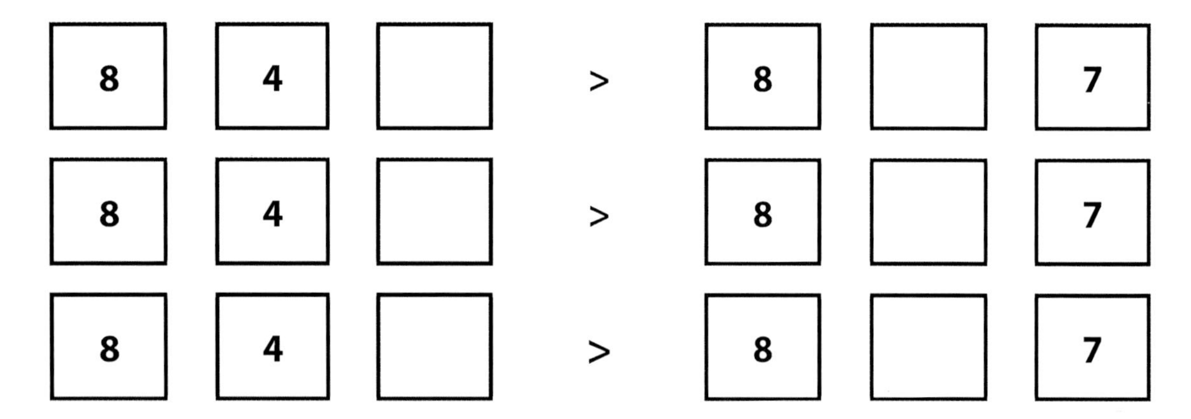

9 I have **6** number cards:

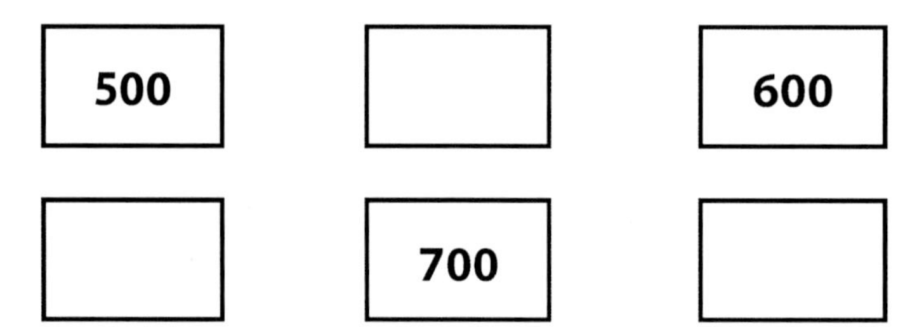

a Choose numbers to write in the empty boxes so that you can make a number sequence. You can move the boxes around.

Write the complete number sequence here:

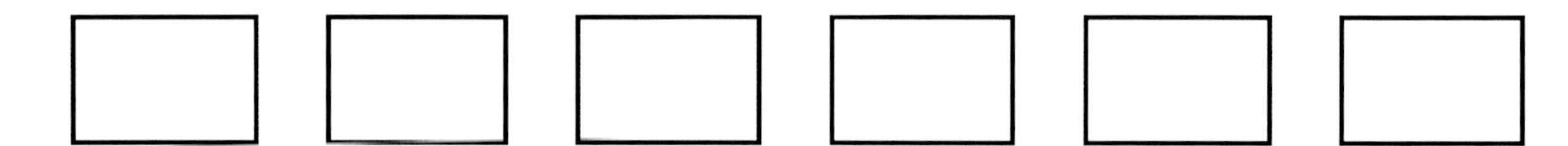

b Choose **three** different numbers so that you make a new sequence. Write your sequence here:

10 Kyle uses counters to make a **3-digit** number.

100s	10s	1s
100 100 100 100 100	10 10 10	1 1 1 1 1 1

a What number did Kyle make? 

b Siddiq made a number with **5** more **tens** than Kyle's number.
Add counters to the diagram above to show Siddiq's number.

c Write Siddiq's number in words.

....

---

11 **a** If you add **3** to a number ending with **8**, you will get a number ending in **1**.

Is this always true, never true or sometimes true?
Explain your answer.

....

....

b What number will be in the ones column if you add **7** to:

678 ☐ **914** ☐ **106** ☐

Name ....

12 **a** Humma thinks of a number.
She adds **100** and then takes away **10**.
Her answer is **495**.

What was her starting number?

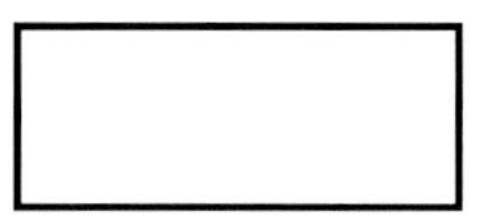

b She thinks of another number.
She adds **8** and then takes away **4**.
Her answer is **342**.

What was her starting number?

13 Join each number to the correct box. One has been done for you:

243	151 → 200
452	201 → 250
198	251 → 300
349	301 → 350
255	351 → 400
423	401 → 450
379	451 → 500

Name ....

14 Add one more number to complete the sequences below.

a

232	332	432	532	

b

736	726	716	706	

c

67	77	87	97	

15 Look at these **3-digit** numbers. Some digits are missing.

Fill the boxes with the same digit so that the numbers are in order from smallest to largest.

Can you do it in **three** different ways?

1		4		8	6	4		4	5	7	

1		4		8	6	4		4	5	7	

1		4		8	6	4		4	5	7	

Name ....

16 **a** Sarah thought of a **2-digit** number between **50** and **100**.

The digit sum was **12**.

The difference between the digits was **2**.

What number did Sarah think of?

b Siddiq thought of a **3-digit** even number between **700** and **800**.

The digit sum was **20**.

What number could Siddiq be thinking of?

Can you find **two** other numbers that would fit?

---

17 Jessica has written down **5** numbers.

The largest number is **482** and the smallest number is **446**.

The digit sum of each number is **14**.

Write down Jessica's numbers from largest to smallest.

482				**446**

Name ....

If you wrote these numbers in order starting with the largest, which number would come **second**?

352 **553** **323** **235** **255** **532**

Explain how you did this.

....

....

a Isiah has written down a **3-digit** number.

The **ones** digit is **half** of the **tens** digit.

The **tens** digit is **half** of the **hundreds** digit.

The **ones** digit is even.

What number did Isiah write down?

b Humma has written down a **3-digit** number.

The **hundreds** digit is **half** of the **ones** digit.

The **ones** digit is **half** of the **tens** digit.

The **hundreds** digit is odd.

What number did Humma write down?

Name ....

20 Draw lines to join each of these numbers to the correct box to show the place value of the digit **2**.

426	**hundreds**
215	**tens**
392	**ones**

---

21 Isiah, Humma and Kyle have picked number cards from a pile. These are the numbers they have picked:

246	**513**	**123**

Isiah's number has a digit sum of **6.**

Humma's number contains consecutive even numbers.

The **hundreds** digit of Kyle's number is **one** less than the **ones** digit of Humma's number.

Can you work out who picked which number?

Isiah [ ] **Humma** [ ] **Kyle** [ ]

Name ....

213 has a digit sum of **6**. **(2 + 1 + 3 = 6)**
Fill in all the boxes below with **3-digit** numbers with a digit sum of **7**.

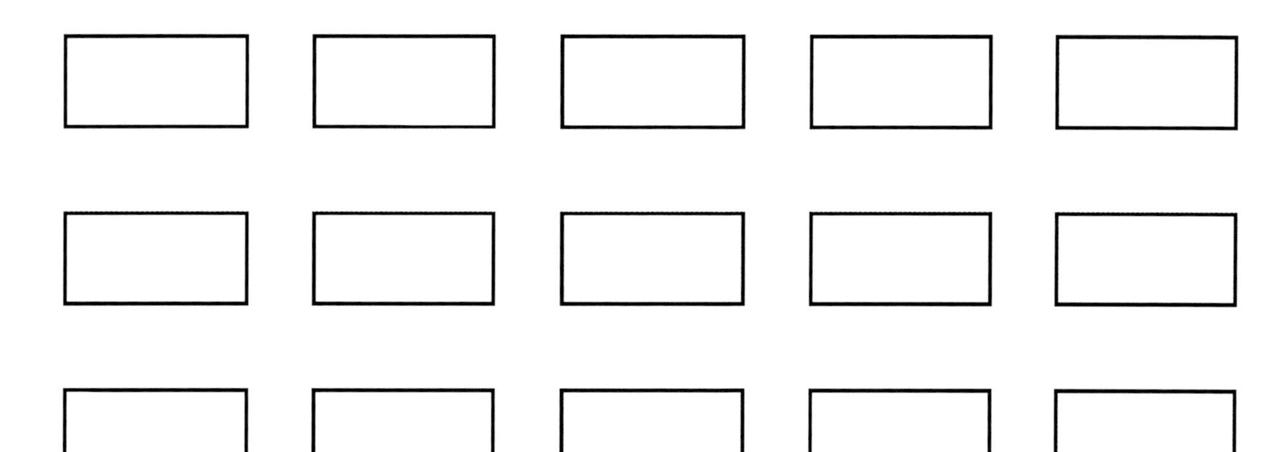

Draw a circle around your **largest** number.

Draw a square around your **smallest** number.

Fill in the missing numbers:

150	200		300	350	

900		700	600		

	750		650		

Name ....

Here is a number line from **0** to **1000**.

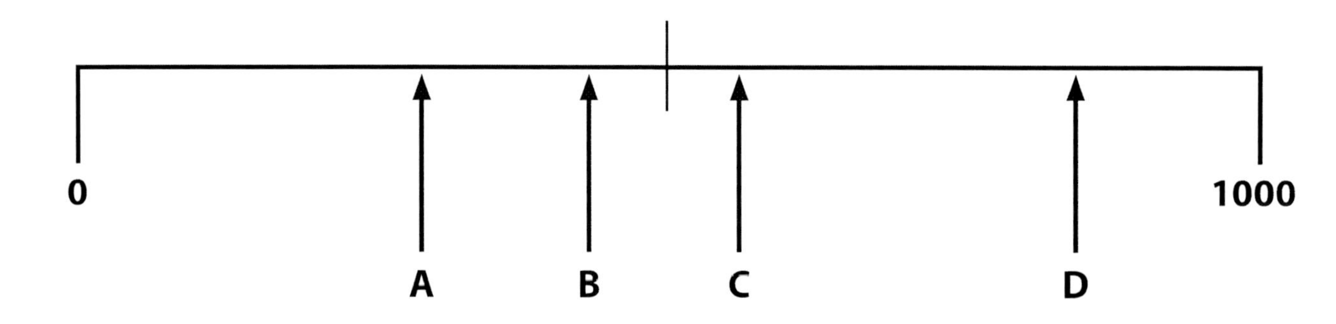

a Which letter is closest to **610**?

b Which two letters are about **250** apart?

c Approximately, what is the difference between **A** and **D**?

Name ....

It may be appropriate for children to use exercise books or paper to record their answers, working out or explanations.

NUMBER

Addition and subtraction

Add a three-digit number and ones (mentally)

1. Isiah has **116** toy cars. For his birthday, he gets another **3**. How many toy cars does Isiah have now?

2. Sarah collects stamps. She has **250** and then gets **9** more. How many does she have now?

3. There are **105** boys and **7** girls in the football tournament. How many children are in the football tournament altogether?

4. Kyle's football team have scored **123** goals so far this season. In the next **two** matches, they score **4** goals and **7** goals. How many goals have they scored altogether?

5. On Saturday, a takeaway sold **134** poppadoms. Kyle then bought **5** and Isiah bought another **7**. How many poppadoms have been sold now?

6. There are **117** grapes, **8** oranges, **9** bananas and **5** melons left in the fruit shop at the end of the day. How many pieces of fruit are left altogether?

Name ....

Subtract a three-digit number and ones (mentally)

1. Sarah has **120** chocolate buttons. She eats **5**. How many buttons does she have left?

2. Humma thought of a number and added **8**. The answer was **128**. What was the number that she first thought of?

3. Isiah has collected **172** football cards. This is **6** more than Kyle. How many football cards has Kyle collected?

4. Sarah collected **147** badges. She gave **9** of them to her friend. How many badges does Sarah have left?

5. Mrs Patel has to drive **248** miles to get home. She drives **9** miles and then stops to buy a newspaper. She then drives another **7** miles and stops to buy petrol. How many more miles does she need to drive to get home?

6. There are **143** apples on the trees. The birds eat **9** and **8** more fall off the trees. How many apples are left on the trees?

Name ....

Add and subtract a three-digit number and ones (mentally)

1. Humma goes to the seaside and collects **120** pretty shells. She gives **9** of them to her mum. How many shells does Humma have left?

2. Jessica builds a tower out of **232** bricks. **7** of the bricks are red and all the others are blue. How many bricks are blue?

3. Siddiq thought of a number, then took away **5**. The answer was **127**. What was the number he thought of?

4. Sarah is at the aquarium. There are **378** little fish in a tank. In the same tank, she counts **9** big orange fish and **6** big blue fish. How many fish are in the tank altogether?

5. Isiah is given **£103** for his birthday. He spends **£8** on a trip to the cinema. Then he buys a new t-shirt for **£9**. How much money does he have left?

£

Name ....

Add a three-digit number and tens (mentally)

1 Sarah can skip **122** times with her skipping rope without stopping. Jessica can skip **20** more times than Sarah without stopping. How many times can Jessica skip?

2 Humma and Sarah go swimming every day for a week. Humma swims **114** lengths and Sarah swims **30** lengths. How many lengths do they swim altogether?

3 In the fruit shop, there are **190** apples and **20** pineapples. How many apples and pineapples are there altogether?

4 There are **17** children on the school bus. At the first stop, **20** children get on and, at the second stop, another **10** children get on. How many children are on the bus now?

5 Sarah was baking for a charity sale. She baked **123** chocolate biscuits, **80** plain biscuits and **40** cupcakes. How many biscuits and cakes did Sarah bake altogether?

Name ....

Subtract a three-digit number and tens (mentally)

1. Kyle has **120** sweets. Humma has **80** sweets. How many more sweets does Kyle have than Humma?

2. In the school hall, the site supervisor puts out **122** chairs for the Christmas Nativity Play. By 6 o'clock, people are sitting on **30** chairs. How many chairs are empty?

3. Mrs Butler's garden has **138** sunflowers, but **50** die because of the bad weather. How many are left?

4. The zookeeper has **212** bananas in a big bag. The monkeys eat **90** bananas for breakfast. How many bananas are left for the monkeys' lunch?

5. Humma gets **104** Christmas cards altogether. She gets **70** from girls. How many cards are from boys?

6. In a cricket match, Isiah's team score **242** runs.
   Humma's team score **90** runs.
   How many more runs did Isiah's team score than Humma's team?

Name ....

Add and subtract a three-digit number and tens (mentally)

1 Jessica scored **102** in a maths test. Kyle scored **90**.
What is the difference between their scores?

2 Mrs Green grew **130** potatoes. She used **20** potatoes to make a shepherd's pie. How many potatoes does she have left?

3 Sarah plants a sunflower and it grows to **120 cm**. Humma's sunflower grows **30 cm** taller. How tall is Humma's sunflower?

cm

4 **143** people went on the school trip this year. This is **thirty** more than last year. How many people went on the school trip last year?

5 Isiah is **90 cm** tall. His brother is **136 cm** tall. How much taller than Isiah is his brother?

cm

6 Kyle has **103** red counters, **30** yellow counters and **80** blue counters. How many counters does Kyle have altogether?

  Name ....

Add a three-digit number and hundreds (mentally)

1. Humma counts **108** red cars and **100** blue cars. How many cars does she count in total?

2. Siddiq wants to build a tower. He has **125** red bricks and **200** blue bricks. How many bricks does he have altogether?

3. A basket has **137** strawberries. Another has **200**. How many strawberries are there altogether?

4. Mrs Bright asked Isiah to work out **138 + 200 + 300**. What answer should he give?

5. Isiah scored **147** runs in a cricket match. Kyle and Humma scored **300** runs between them. How many runs did they score altogether?

6. There are **178** sheep, **200** pigs, **100** cows and **400** chickens on the farm. How many animals are there on the farm altogether?

Name ....

Subtract a three-digit number and hundreds (mentally)

1. There are **350** pupils at Sunshine Primary School. Classes 3, 4 and 5 go on a trip to the theme park, leaving **200** children at school. How many children went on the trip to the theme park?

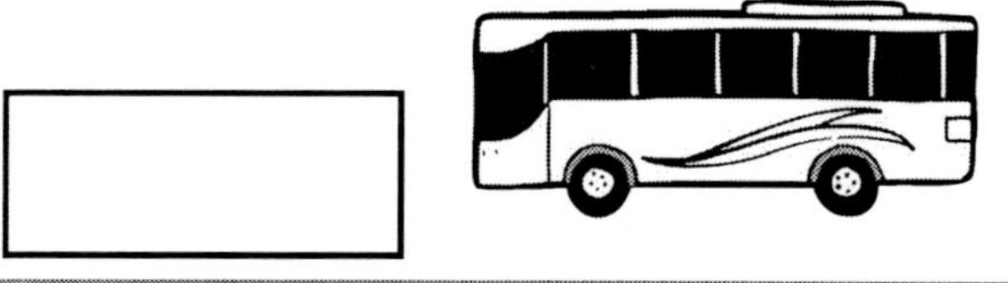

2. Humma's team score **180** runs in a cricket match. Jessica's team score **100** runs. How many more runs do Humma's team score?

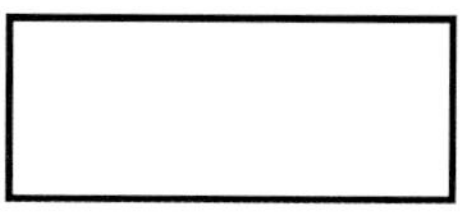

3. Sarah is reading a book. The book has **357** pages. She reads **200** pages. How many pages does she have left to read?

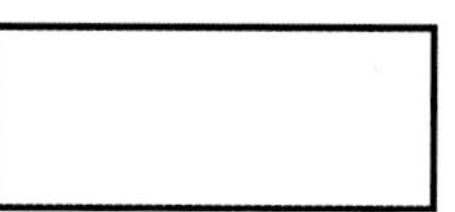

4. There are **180** bananas and **200** apples for the children at break. **300** pieces of fruit are eaten. How much fruit is left over?

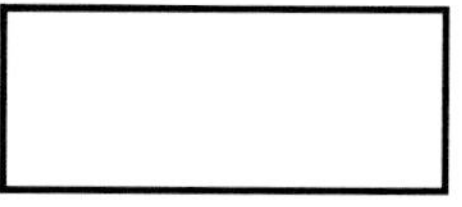

5. There are **917** fiction books on the shelves in the library. On Monday, children take away **200** to read and, on Wednesday, children take away **400** to read. How many books are left on the shelves?

The Lion, the Witch and the Cupboard

Name ....

Add and subtract a three-digit number and hundreds (mentally)

1 Between 9 am and 5 pm a café sold **200** cups of coffee and **145** cups of tea. How many drinks did they sell altogether?

2 There are **328** sweets in a tub. Humma and Isiah eat **200** of the sweets. How many sweets are left?

3 Siddiq and his grandad are making a cake. They start by putting **175 g** of sugar and **300 g** of butter into a bowl. How many **grams** of sugar and butter are in the bowl?

g

4 Sarah is making necklaces. She has **762** beads. She uses **100** beads to make a necklace for her mum and **300** beads to make a necklace for her sister. How many beads does she have left?

5 Kyle has **756** trading cards. Isiah has **300** cards and Jessica has **100**. How many more trading cards does Kyle have than Isiah and Jessica put together?

Name ....

Add numbers with up to three digits using a formal written method

(If you need more space, use your book or paper.)

1 Complete the written method that Kyle might use to calculate **42** add **88**.

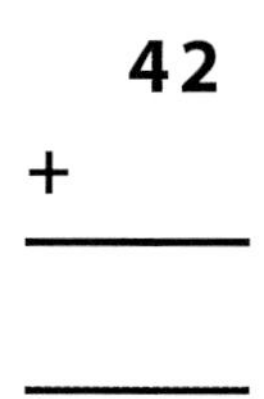

2 Sarah collected **97** badges. Her mum bought her **45** more badges. How many badges does she have now?

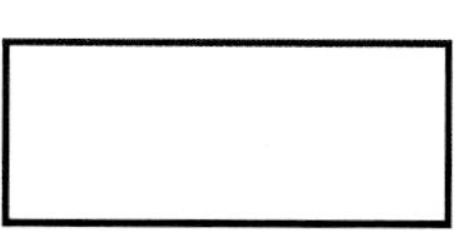

3 There are **336** children and **45** staff at St Michael's Primary. How many people are in the school altogether?

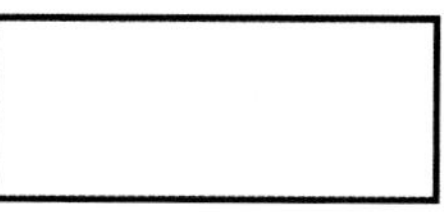

4 Miss Thompson wrote this problem on the whiteboard. Use a written method of column addition to solve it.

234 + 87 + 9 =

Name ....

Subtract numbers with up to three digits using a formal written method

(If you need more space, use your book or paper.)

1. Humma has **£46** and she buys a new dress for **£12**. How much does she have left to buy some shoes? Set out your written method carefully and work out the answer.

£ ☐

2. Show how Jessica might use a written method to calculate **96** subtract **32**. What is her answer?

☐

3. Isiah solves **85** subtract **37** using a written method. How might he do it? What would his answer be?

☐

4. How might Sarah use a written method to subtract **48** from **396**? Show how she would do this and then write the answer.

☐

5. Mr Berry writes this problem on the whiteboard. Use a written method of column subtraction to solve it.

505 – 166 =

☐

Name ....

Add and subtract numbers with up to three digits using a formal written method

(If you need more space, use your book or paper.)

1. Isiah collected **265** conkers. He gave **52** of them to his friend. How many conkers did Isiah have left? Use a written method to work out the answer.

2. How would Kyle use a written method to calculate **455** subtract **224**? What would his answer be?

3. Siddiq used a written method to find the total of **154** and **286**. Show how you think he did it and then write the answer.

4. Humma and Siddiq are members of a skiing club. Humma has been a member for **657** days and Siddiq has been a member for **814** days. How many more days has Siddiq been a member than Humma? Use a written method.

5. Jessica is **126 cm** tall. Sarah is **117 cm** tall. Humma is **137 cm** tall. What is the total of their heights?

cm

Name ....

Estimate the answer to a calculation

1 Sarah estimates that **11 + 18** would be about **30**. Is this a good estimation? Explain your answer.

....

....

2 Mrs Stone asks Jessica to estimate the answer to **48 + 52**. Circle the best estimation.

40 + 50 **50 + 50** **50 + 60**

3 Mr Malik buys **4** packs of sweets. Each pack has **10** sweets. There are **36** children in his class. Will he have enough sweets to give to each child? Explain your answer.

....

....

4 There are **409** people at the cinema. If **311** people leave early, estimate, to the nearest **hundred**, how many people are left.

5 Humma is asked to subtract **196** from **303**. Use rounding to estimate the answer for Humma.

6 After rounding to the nearest **100**, Miss Bradley says there are **500** children in the school. What is the least number of children there could be in school?

Name ....

Use inverse operations to check answers

1 Can you help Siddiq turn this number statement into a subtraction?

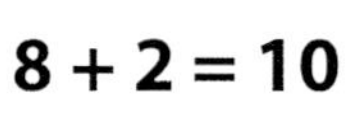

8 + 2 = 10

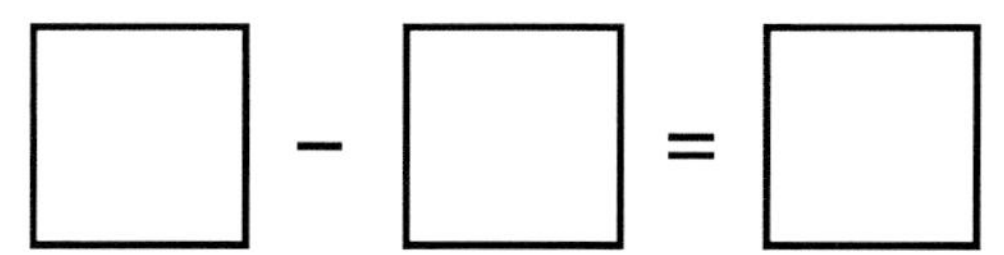

☐ – ☐ = ☐

2 Kyle says, "I can check my answer to **4 + 8** = **12** by subtracting **12** from **8**." Is Kyle correct?
Explain your answer.

Yes / No

....

....

3 Mr Mitchell wrote these calculations on the board. Can you fill in the missing numbers?

☐ + **8** = ☐ **15 – 8 = 7**

4 Isiah works out that **9 + 7** = **16**. How could he check her answer using subtraction?

☐

5 Sarah writes an inverse of **32 + 80** = **112**. What number statement might she write?

☐

6 Humma checks her answer to **97 – 78** = **19** using the inverse. Show how she would do this?

☐

Name ....

Solve problems, including missing number problems, using number facts and place value

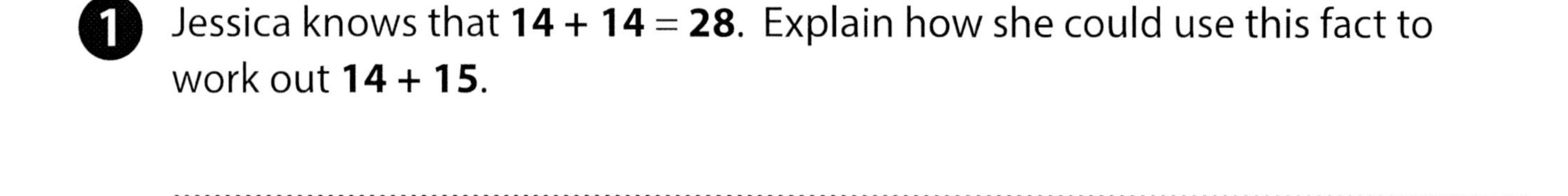

1 Jessica knows that **14 + 14** = **28**. Explain how she could use this fact to work out **14 + 15**.

....

....

2 If **half** of **36** is **18**, what is **double 18**?

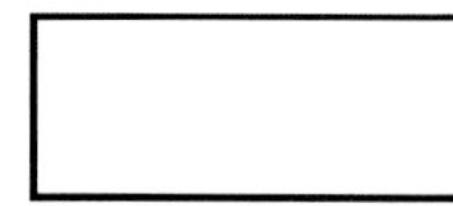

3 Kyle has a secret number. It is less than **20**, has **2** digits and is a multiple of **4**. The digits add up to **7**. What is Kyle's number?

4 Sarah is selling her home-made orange juice. She squeezes **228** oranges on Monday, **193** on Tuesday and **277** on Wednesday. How many oranges has she used altogether?

5 Humma is trying to fill in the number statements by putting a number on each line. Can you help her complete the missing numbers?

☐ − **8** = **12**

21 − ☐ = **13**

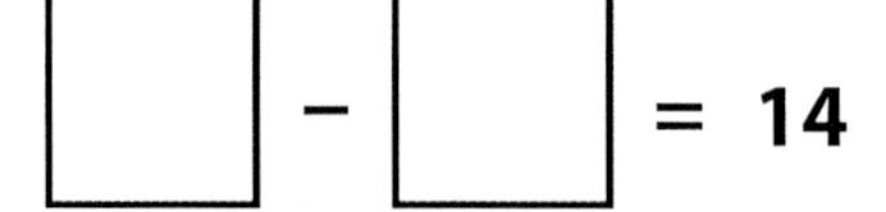

☐ − ☐ = **14**

Name ....

Solve addition money problems

1 On Saturday, Isiah's mum gives him **52p** and his dad gives him **45p**. How much does Isiah get altogether?

[ ] p

---

2 It costs **£1.50** to go swimming. How much would it cost Siddiq and Sarah to go swimming together?

£ [ ]

---

3 Mum buys Jessica and Sarah a pizza each. **One** pizza costs **£2.50**. How much money does mum spend?

£ [ ]

---

4 Humma buys a comic for **£1.50** and some new hair clips for **99p**. How much money does she spend in total?

£ [ ]

Look at the cost of sweets in the shop and then answer Questions 5 and 6.

Chewys – 17p	**Fruit Pops – 11p**	**Swirlys – 9p**
Frizzles – 15p	**Toffee Treats – 25p**	

---

5 Sarah buys a Frizzle, a Swirly and a Fruit Pop. How much does she spend?

[ ] p

---

6 Jessica has **50p**. Has she enough to buy **two** Frizzles and **one** Toffee Treat? Explain your answer.

Yes / No

....

....

Name ....

Solve subtraction money problems

1 How much change does Humma get from **67p** if she buys an apple from the fruit shop for **25p**?

[ ] p

2 Sarah has **£3**. She buys a gingerbread man for **75p**. How much money does she have left?

£ [ ]

3 Kyle has saved **£12**. He spends **£7.25** on a DVD. How much money does he have left?

£ [ ]

4 For Eid, Siddiq gets **£110**. He buys a game costing **£38**. How much will he have left?

£ [ ]

5 Jessica has **£192**. She spends **£56** at the toy shop. Then she goes to the clothes shop and spends **£73**. How much does Jessica have left?

£ [ ]

6 Isiah gets **£163** from all his family for his birthday. He buys a game for **£37** and a DVD Boxset for **£18**. He also spends **£48** on a party. How much does he have left?

£ [ ]

Name ....

Solve addition and subtraction problems

1. There are **37** blue soft balls and **53** red soft balls in a play area. How many soft balls are there altogether in the play area?

2. Sarah has **62p**. She gives Sididq **25p**. How much does Sarah have left?

   p

3. In the Sandwich Shop, there are **13** cheese and tomato sandwiches, **25** turkey sandwiches and **14** ham sandwiches. Kyle buys **2** turkey sandwiches and **1** ham sandwich. How many sandwiches are left in the shop?

4. In the PE hall, there are **275** bean bags. Mrs Clayton's class takes **115**, Mr Thabani's class takes **94** and Mrs England's class **30**. How many bean bags are left?

5. Kyle takes **86** oatmeal biscuits, **42** chocolate biscuits and **27** ginger biscuits to a party. At the end of the party, there are **39** biscuits left. How many biscuits have been eaten at the party?

6. Sarah thinks of a number. She adds **20** and subtracts **152**. The answer is **83**. What was her number?

Name ....

Solve addition and subtraction problems

1 Humma thinks of a number then subtracts **15**. The answer is **30**. What was her number?

---

2 Kyle has **88p** and Siddiq has **47p**. What is the difference in the amount of money that they have?

p

---

3 Isiah scores **240** on a computer game. This is **82** more than Kyle's score. What was Kyle's score?

---

4 Mrs Anderson bought **250** bottles of water for her family. In June, they drank **57** bottles. In July, they drank **84** and in August, they drank **79**. How many bottles were left?

---

5 Class 3 recorded how many apples were sold in the school shop. Look at their results for week one.

Monday	Tuesday	Wednesday	Thursday	Friday
16	19	21	24	26

During week two, the shop sold **20** fewer apples than during week one. How many apples were sold in week two?

Name ....

Solve addition and subtraction problems

1 Kyle's favourite team has **48** points and his brother's favourite team has **39** points. How many points do the teams have in total?

---

2 Sarah is saving up to buy some earrings for her mum's birthday. The earrings cost **£29**. Sarah has only **£12**. How much more money does she need?

£

---

3 There are **258** people in the queue for the Big Bouncy Rollercoaster. **62** people leave the queue early because they are hungry. How many are still in the queue?

---

4 Isiah thinks of a number, **doubles** it and adds **4**. His answer is **44**. What was Isiah's number?

---

5 Sarah solves **984** minus **368** using a written method. What would her answer be?

---

6 On a farm, there are **340** chickens in a chicken coop. The farmer puts another **114** chickens into the coop. The farmer then takes **98** chickens from the coop to sell at the market.
How many chickens are left?

Name ....

It may be appropriate for children to use exercise books or paper to record their answers, working out or explanations.

MASTERING

Addition and subtraction

1 Sarah has completed some calculations.
Unfortunately, she has made mistakes in all of them.
Check each calculation and write the correct answers where necessary.
Remember to show all your working.

a

	4	7	5
+	3	8	7
	7	5	2

b

	5	0	6
−	2	0	7
	2	5	9

c

	3	9	5
+	2	9	8
	6	9	2

d

	5	2	4
−	2	2	7
	3	2	3

2 Humma has **4** different coins in her pocket.

a What is the largest amount of money she could have?

£

b What is the least amount of money she could have?

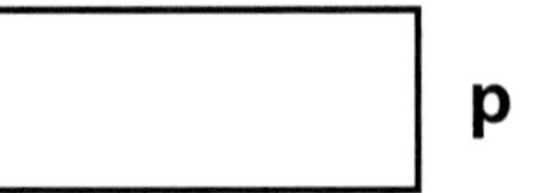

p

c What is the difference between the largest and smallest amounts?

£

Name ....

3 This model shows related number facts.

480	
256	224

Write down all the related number facts that the model shows.

480	–	256	=	
	+		=	
	–		=	
	+		=	

4 Decide whether these statements are always true, sometimes true or never true. For each one, give **two** examples to explain your answer.

statement	always, sometimes, never	examples
The sum of two even numbers is even.		
The sum of two odd numbers is odd.		
The sum of three odd numbers is odd.		

Name ....

5 **a** Siddiq says, "The sum of **three** even numbers is always even."

Is he correct? **Yes / No**

Explain your answer.

....

....

b Humma says, "The sum of **three** odd numbers is always odd."

Is she correct? **Yes / No**

Explain your answer.

....

....

---

6 Isiah spends **35p**. Sarah spends **62p**.

How much more has Sarah spent than Isiah?

[ ] **p**

Humma says, "To work this out you should do **35 + 62**."

Is Humma correct? **Yes / No**

Explain your answer.

....

....

Name ....

7 How many ways you can make **50p** using just silver coins?

How do you know that you have found all the solutions?

....

....

8 Siddiq has read **154** pages of a book.
Kyle has read **25** pages more than Siddiq.
How many pages has Kyle read? [ ]

Sarah says you need to do the calculation **154 + 25**.
Humma says you need to do the calculation **154 – 25**.
Who is correct? **Sarah / Humma**

Explain your answer.

....

....

Name ....

9 Use **single-digit** numbers to complete this calculation.
Can you find 14 different ways?

☐ + ☐ + ☐ = **15** ☐ + ☐ + ☐ = **15**

☐ + ☐ + ☐ = **15** ☐ + ☐ + ☐ = **15**

☐ + ☐ + ☐ = **15** ☐ + ☐ + ☐ = **15**

☐ + ☐ + ☐ = **15** ☐ + ☐ + ☐ = **15**

☐ + ☐ + ☐ = **15** ☐ + ☐ + ☐ = **15**

☐ + ☐ + ☐ = **15** ☐ + ☐ + ☐ = **15**

☐ + ☐ + ☐ = **15** ☐ + ☐ + ☐ = **15**

---

10 Isiah has **£50**.
He buys a book for **£9** and a pen for **£14**.
How much money does Isiah have left?

£ ☐

Name ....

11 Jessica found she could add **three 2-digit** numbers to make **171**.
Unfortunately, she has forgotten what the numbers were.
She does remember that she used only the digits **2** and **7**.
Can you discover Jessica's calculation?

☐☐ + ☐☐ + ☐☐ = **171**

12 Fill in the boxes to make the calculation correct.

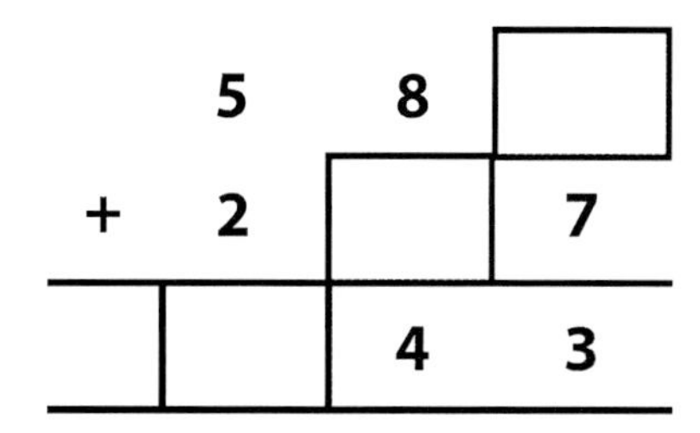

13 Use the symbols **+** or **–** to make these number sentences correct.

24 ☐ **76** = **100**

56 ☐ **29** = **27**

80 ☐ **35** = **45**

46 ☐ **28** = **74**

500 ☐ **350** = **850**

Name ....

14 Siddiq says, "**75 + 60 = 135**. This means that the total of **two 2-digit** numbers is always over **100**"

Is he correct? **Yes / No**

Explain your answer.

....

....

15 Add one digit to each box to make these calculations correct.
Can you do it 7 different ways?

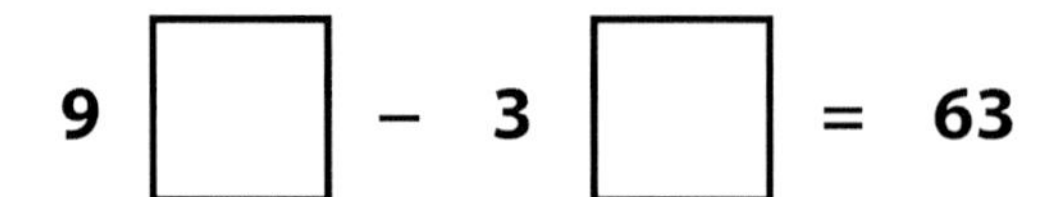

9 ☐ **–** **3** ☐ **=** **63**

9 ☐ **–** **3** ☐ **=** **63**

9 ☐ **–** **3** ☐ **=** **63**

9 ☐ **–** **3** ☐ **=** **63**

9 ☐ **–** **3** ☐ **=** **63**

9 ☐ **–** **3** ☐ **=** **63**

How do you know if you have found all the solutions?

....

....

Name ....

16 Isiah says that if you add **6** to a number ending in **7**, the ones digit will always be **3**.

Is Isiah correct? **Yes / No**

Explain your answer.

....

....

17 Use lines to join these calculations up to the best estimation of the answer.

17 + 52	**30**
102 – 68	**40**
578 – 517	**50**
22 + 19	**60**
249 – 205	**70**

18 Sarah knows that **146 + 30 = 176**.

Complete these number sentences using this fact to help you.

☐ + **146** = **176**

176 – **30** = ☐

☐ – **146** = **30**

Name ....

19 Humma, Sarah and Jessica go to buy some new stationery for school.

Item	Cost
pencil	**26p**
rubber	**34p**
sharpener	**53p**
ruler	**47p**

a Humma buys **2** pencils and an rubber.
How much change does she get from **£2**?

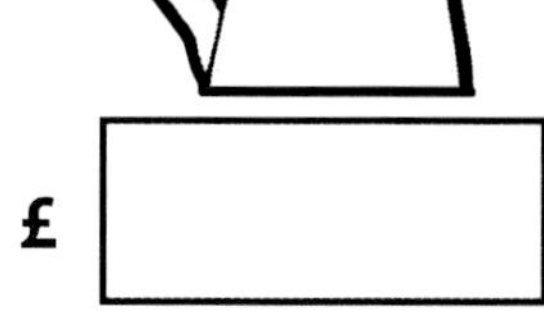

b Sarah buys **1** of each item on sale.
How much more does Sarah spend than Humma?

p

c Jessica spends exactly **105p**.
Which three items did Jessica buy?

20 The addition pyramid is made by adding **two** numbers in the bottom row to make the number in the row above.

a Complete this addition pyramid.

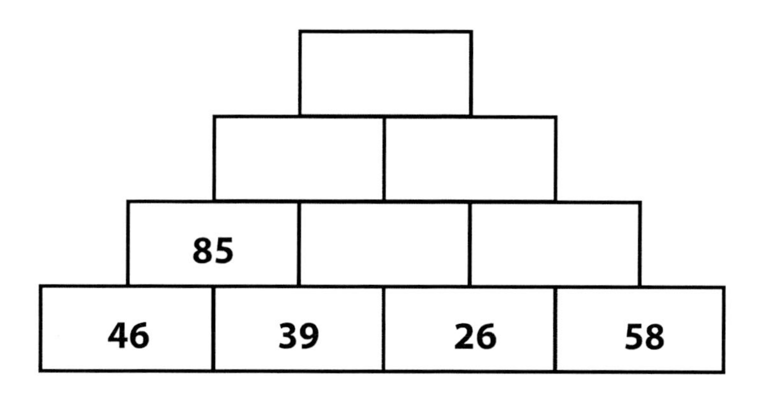

b Find the missing numbers in this addition pyramid. You may need to use subtraction.

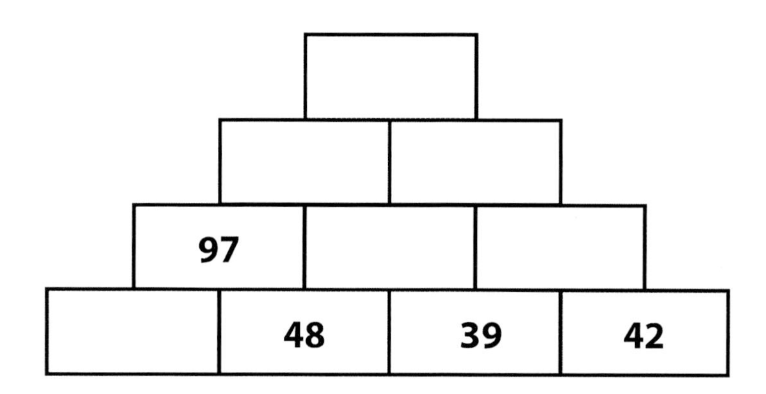

c Find the missing numbers in this addition pyramid. You may need to use subtraction.

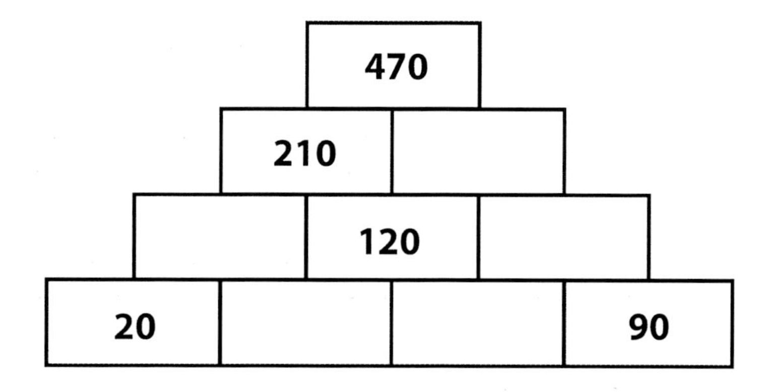

Name ....

21 The table shows the number of children who played football, netball or hockey during one week.

	Monday	Tuesday	Wednesday	Thursday	Friday
football	27	43	36	34	25
netball	72	15	68	46	52
hockey	69	37	49	57	77

a) How many children played football during the week?

b) How many more children played hockey on Friday than played netball on Tuesday?

c) How many children played some sport on Wednesday?

d) Which was the most popular sport?

e) Which was the least popular sport?

Name ....

Siddiq and Isiah have 46 Top Trumps cards.
Siddiq has 25 cards.

a How many cards does Isiah have?

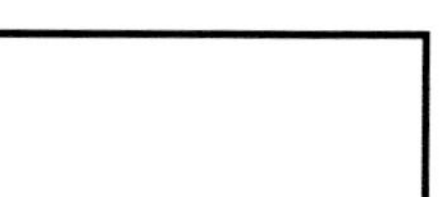

b Isiah and Kyle have 37 cards.
How many cards does Kyle have?

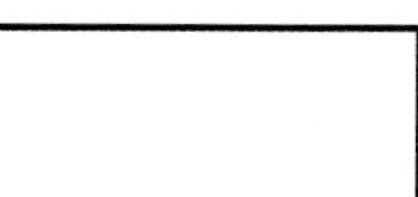

c Siddiq gives 3 cards to Isiah and 7 cards to Kyle.
How many cards do they each have now?

Siddiq ☐ **Isiah** 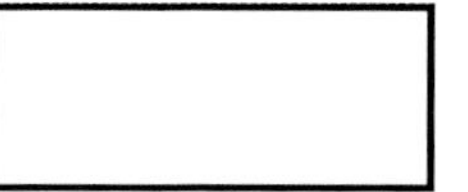 **Kyle** ☐

What are the missing digits in this calculation?

2 ☐ **6** + **3** **5** ☐ = **600**

24 The table shows the prices of food and drink from a local café.

Main meals	Desserts	Drinks
fish and chips £2.35	apple pie £1.75	cola 55p
chicken salad £2.15	chocolate cake £1.45	milk 40p
tuna pasta £2.20	ice cream £1.30	water 30p

a Kyle buys a chicken salad, ice cream and milk.
How much change does Kyle get from **£5?**

£ ☐

b Sarah has **£3.75** to spend.
She wants to buy a main meal, a dessert and a drink.

Does she have enough money? **Yes / No**

Which items could she afford to buy?

Main meal ....

Dessert ....

Drink ....

It may be appropriate for children to use exercise books or paper to record their answers, working out or explanations.

NUMBER

Multiplication and division

Recall and use multiplication facts for the 3 times table

1. It costs **£3** to go swimming. How much would it cost for **4** children to go swimming?

£ ☐

2. There are **5** apples in a bag. How many apples would there be in **3** bags?

☐

3. A toy car has **four** wheels. How many wheels would **three** toy cars have?

☐

4. Kyle can run **7** miles in an hour. How many miles could he run in **3** hours?

☐ **miles**

5. Siddiq has **3** toy sets with **10** dinosaurs in each set. How many dinosaurs are there altogether?

☐

6. Isiah got **80p** each week for his spending money. He saved this money for **3** weeks. How much did Isiah save altogether?

Isiah

£ ☐

Name ....

Recall and use multiplication facts for the 4 times table

1. **Four** football cards come in a pack. Kyle buys **two** packs. How many cards does he have?

2. The puppy eats **3** Doggy Dinners every day. How many Doggy Dinners does he eat in **four** days?

3. Taxis hold **four** people. How many people would **4** taxis hold?

4. There are **10** carrots in a row in Mr Grime's garden. How many carrots are there in **4** rows?

5. Isiah has a pack of **seven** colouring pencils. How many colouring pencils would he have if he had **four** packs?

6. Mr Fisher gives out **four** sweets to each child. He gives sweets out to **12** children. How many sweets does he give out altogether?

Name ....

Recall and use multiplication facts for the 8 times table

1. There are **8** biscuits in a pack. Mrs Piper buys **two** packs. How many biscuits has she bought?

2. Sarah has **3** pots of daffodils. Each pot has **8** daffodils growing in it. How many daffodils are growing altogether?

3. Jessica fills **6** boxes with **8** cupcakes in each box. How many cupcakes are there altogether?

4. In a netball competition, there are **eight** teams. Each team has **seven** players. How many players are there in total?

5. **One** spider has **8** legs. Siddiq has **8** pet spiders. How many legs do the spiders have altogether?

6. Humma likes to read **8** books a month. How many books would she read in a year?

Name ....

Recall and use multiplication facts for the 3, 4 and 8 times tables

1. There are **8** pens in a pack. Sarah buys **2** packs. How many pens does she buy altogether?

2. Kyle is making dinner. He puts **3** potatoes on each plate. How many potatoes would he need for **4** plates?

3. Giraffes have **4** legs. There are **6** giraffes at the safari park. How many giraffe legs are there altogether?

4. There are **8** cereal bars in a packet. How many cereal bars do you have if you buy **8** packets?

5. Sarah rides her horse, Harry, **four** times a week. How many times does she go horse riding in **11** weeks?

6. Kyle, Humma and Isiah have **12** football cards each. How many cards do they have altogether?

Name ....

Recall and use multiplication facts for the 3, 4 and 8 times tables

1. **One** bag holds **4** tomatoes. How many tomatoes are in **2** bags?

2. Siddiq builds a tower out of **8** bricks. How many bricks would he need to build **4** towers?

3. Mr Fowler eats chicken for his dinner **three** times a week. How many times does he eat chicken in **ten** weeks?

4. Mrs Edwards splits her class into **8** groups. Each group has **3** children. How many children are in her class?

5. At the One-Stop Pet Shop, there are **12** spiders. Each spider has **8** legs. How many spider legs are there altogether?

6. The hen lays **3** eggs per week. Another hen lays **4** eggs per week. In **7** weeks, how many eggs do they lay altogether?

Name ....

Recall and use division facts for the 3 times table

1. Sarah cooks **6** samosas. Sarah and her brother eat the same amount of samosas and they eat all of them. How many samosas do they each eat?

2. There are **18** chocolates in a box. How many people can have **3** chocolates each?

3. There are **15** strawberries in a basket. Mr Murray shares them equally between his **3** children. How many strawberries do they each have?

4. The baker puts **3** chocolate drops on each cake. He uses **24** chocolate drops. How many cakes has he baked?

5. Mrs Grant has **3** daughters. She buys **27** bags of beads for necklace making and shares them equally between her daughters. How many bags of beads do her daughters each have?

6. Kyle has **36** mini eggs. He shares them equally between himself and his friends, Siddiq and Jessica. How many mini eggs will they each get?

Name ....

Recall and use division facts for the 4 times table

1. Sarah cuts her birthday cake into **8** pieces. She shares it between **4** friends. How many pieces of cake will they each have?

2. There are **12** biscuits in a packet. How many biscuits would **4** people get each?

3. Humma has **16** stickers to put in her sticker book. She can put **4** stickers on each page. How many pages will she use?

4. Kyle has **24** lemons. **4** lemons make a glass of lemonade. How many glasses of lemonade can Kyle make?

5. Ms Khan has **32** reading books in her class. She puts **4** books on each table. How many tables are in her class?

6. There are **48** children on the playground. Mr White puts the children in **4** equal lines. How many children are in each line? If **4** more children came onto the playground, how many would be in each line then?

Name ....

Recall and use division facts for the 8 times table

1. Humma picks **16** daisies to make daisy chains. She makes each daisy chain with **8** daisies. How many daisy chains can she make?

2. There are **32** children in Year 3. The teacher divides his class into **8** groups. How many children are in each group?

3. **56** cars are in a car park. There are **8** cars on each level. How many levels are in the car park?

4. Miss Harper picked **64** apples from the trees in her orchard. She put **8** apples in each basket. How many baskets did she need?

5. A fruit stall has **80** bananas. There are **10** bananas in each bunch. How many bunches of bananas are there?

6. The pet shop has **88** goldfish. There are **8** fish in each tank. How many fish tanks are there? How many fish would there be in **12** tanks?

Name ....

Recall and use division facts for the 3, 4 and 8 times tables

1. There are **9** ice pops in a box. **3** children share them equally. How many ice pops do they each have?

2. There are **16** bouncy balls in a box. They are given out to **8** children. How many bouncy balls does each child get?

3. Mrs Gardener grew **18** pumpkins. There are **3** pumpkins in each row. How many rows are there?

4. There are **32** roses in Mr Locke's garden. He gives his wife a bunch of **8** roses. How many more bunches of **8** roses can he make?

5. Sarah needs to cook **48** pies. She puts **4** pies on each tray to cook. How many trays of pies will there be?

6. Isiah has **72** strawberries. To make **one** milkshake, he needs **8** strawberries. How many strawberry milkshakes could he make with all the strawberries?

Name ....

Recall and use division facts for the 3, 4 and 8 times tables

1 There were **12** dolls in the toy box. The dolls were shared between **4** children. How many dolls did they each have?

2 **Three** eggs are needed to make a cake. Kyle has **twelve** eggs. How many cakes could he make?

3 Mr Akhtar shares **£28** between his **4** sons. How much money do they each get?

£

4 Sarah bakes **36** jam tarts for her party guests. She gives them **3** each. How many guests are given jam tarts?

5 Jessica has **72** printed photographs. She puts them into a photo album. Each page holds **8** photographs. How many pages does she use?

6 Kyle, Siddiq, Jessica and Humma share **48** sweets. How many sweets do they each get?

Name ....

Recall and use multiplication and division facts for the 3, 4 and 8 times tables

1. Mrs Lawson drives her car **3** times a day. How many times does she drive her car in **4** days?

2. Mum bought **20** cartons of apple juice. There are **4** cartons in a box. How many boxes did mum buy?

3. If **24** children are divided into **three** teams, how many children are in each team?

4. Kyle is decorating cakes. He has **36** jelly sweets. He puts **4** sweets on each cake. How many cakes can he decorate?

5. Humma saves **64** pounds for her holiday to Scotland. She divides the money equally between the **8** days she is there. How much money does she have for each day?

   £

6. Jessica thinks of a number and multiplies it by **4**. The answer is **52**. What was her number?

Name ....

Recall and use multiplication and division facts for the 3, 4 and 8 times tables

1. Siddiq has **3** tubes of sweets. Each tube has **8** sweets. How many sweets does he have?

2. Kyle is collecting money for charity. He collects **£4** each day for **10** days. How much money does he collect altogether?

   £

3. Year 3 planted poppies in the school garden. They planted **6** rows of **8** seeds. How many poppies did Year 3 plant?

4. **28** biscuits are shared equally between **four** tables in Miss Thompson's class. How many biscuits are put on each table?

5. Isiah is saving up for some new trainers. He saves **£3** each week for **12** weeks. How much money does Isiah save?

   £

6. Siddiq has **36** football cards. He puts them into **6** equal piles. How many football cards does he put in each pile?

Name ....

Solve problems involving doubling and connecting the 2, 4 and 8 times tables

1 Kyle knows his **2** times table but not his **4** times table. Explain, using numbers or words, how he can solve **4 x 4** using his **2** times table.

....

....

2 Sarah knows that **2 x 4** is **8**. How could she use this to help her work out **2 x 8**? Explain your answer using numbers or words.

....

....

3 There are **8** people on the bus. **8** more people get on the bus. How many people are on the bus altogether? Work this out using the **2** times tables.

4 Mr Khan has banned the **8** times table in his class. He asks Sarah to work out **8 x 8** by using the **4** times table. Explain how she could do this.

....

....

5 Jessica has **6** tubs, each containing **8** buttons. How many buttons does she have altogether? Use your knowledge of the **4** times table to help you. Explain how you found your answer.

....

....

Name ....

Solve problems involving multiplication of a two-digit number by a one-digit number, using a mental method

1. A box of chocolates has **12** chocolates in each layer. There are **4** layers in a box. How many chocolates are there in the box altogether?

2. Isiah swims **8** lengths of a **10 metre** pool. How many **metres** does he swim in total?

   m

3. Jessica buys **5** apples for **24p** each. How much does she spend?

   £

4. Humma has **23** friends coming to her birthday party and she wants to make sure they have **3** cupcakes each. How many cupcakes will she need to bake?

5. There are **44** people on the aeroplane. Each person has **3** items of luggage. How many items of luggage are there altogether?

6. Sarah has **4** jars with **53** marbles in each jar. How many marbles does she have altogether?

Name ....

Solve problems involving multiplication of a two-digit number by a one-digit number, using a mental method

1. In assembly, there are **8** children in each row. There are **12** rows. How many children are there in assembly?

2. Kyle is building Lego towers. To build **one** tower, he needs **38** pieces of Lego. How many pieces does he need to build **2** Lego towers?

3. If **30** children have **five** reading books each, how many books do they have altogether?

4. Jessica's sister got **£26** each week from working at the café on Saturday. She saved this money for **4** weeks. How much did she save?

   £

5. **27** children each put **five** Christmas cards into the Christmas post box. How many cards were in the post box altogether?

6. Dad has been saving coins. He has **6** jars with **28** coins in each jar. How many coins does he have in total?

Name ....

Solve problems involving division of a two-digit number by a one-digit number, using a mental method

1. There are **72** grapes in a bag. How many people can have **8** grapes each?

2. Siddiq has **24** t-shirts. His mum asks him to share them equally between **3** drawers in his dressing table. How many t-shirts does he put in each drawer?

3. Sarah has **60p**. She shares it equally between herself, Jessica and Siddiq. How much do they each get?

   p

4. A big box of biscuits has **75** biscuits in **5** layers. How many biscuits are in each layer?

5. Humma thinks of a number and multiplies it by **4**. Her answer is **160**. What was her number?

6. Mum puts **£45** in a tin and Dad adds another **£20**. They share the money equally between their **five** children. How much does each child get?

   £

Name ....

Solve problems involving division of a two-digit number by a one-digit number, using a mental method

1. In Miss Khan's class, there are **28** children. She divides her class into **2** groups. How many children are in each group?

2. A taxi can carry **4** people. How many taxis would be needed to take **32** people to the One Destination music concert?

3. Perry's Pet Shop keeps **5** stick insects in each tank. How many tanks would be needed for **75** stick insects?

4. There are **96** bananas for break time. **3** classes each get an equal amount. How many bananas does each class get?

5. At the theme park, the Tummy Tickler ride takes **8** people at a time. How many rides are needed so that **96** people can have a turn?

6. Kyle has **96** sweets, shared equally between **4** bags. If he eats **3** sweets from each bag, how many sweets would be in each bag then?

Name ....

Solve problems involving multiplication using a formal written method

(For each of the problems show the written method in your book or on paper.)

1. In the furniture store, each table has **4** legs. How many legs would **22** tables have?

2. Miss Jones asks Siddiq to work out **24** multiplied by **3**. Show how he would do this.

3. There are **52** cards in a pack. How many cards are there in **5** packs?

4. In the theatre, there are **8** rows of **17** chairs. How many chairs are there altogether in the theatre?

5. One Tongue Twister sweet costs **16p**. Isiah buys **one** for himself and his **7** friends. How much does he spend?

£

6. Which is more: **24** multiplied by **8** or **46** multiplied by **4**? Show your written multiplication method for both calculations.

Name ....

Solve problems involving division using a formal written method

(For each of the problems show the written method in your book or on paper.)

1. How many teams of **4** can be made from **84** people?

2. A group of **56** children are going to the fair. They are travelling in cars. **4** children can fit into each car. How many cars are needed for the trip?

3. A pizza serves **3** hungry children. How many pizzas would be needed to feed **72** hungry children?

4. The miniature steam train holds **8** people at a time for a circuit of the track. How many times would the train have to go round the track if **96** people wanted a ride?

5. Isiah is giving out coloured pencils. He gives **8** pencils to each person. Isiah has **99** pencils to give out. How many people are given pencils? How many pencils would be left over?

people **left over**

Name ....

Solve problems involving multiplication and division using a formal written method

(For each of the problems show the written method in your book or on paper.)

1. Mr Patel gives each of his **four** sons **£16**. How much does he give to his sons altogether?

£ ☐

2. How many teams of **3** can be made from **39** people?

☐

3. In the PE cupboard, there are **94** bean bags. Ms Bluebell gives out the bean bags to **8** groups. How many bean bags does each group get? How many bean bags are left over?

☐ ☐

4. Use a formal written method to work out **26** multiplied by **3**.

☐

5. Humma swims **46** lengths a week. How many lengths does she swim in **8** weeks?

☐

6. In a week at the zoo, **97** fish are shared betweeen **3** penguins. There are some fish left over. How many fish does each penguin get and how many are left over?

☐ ☐

Name ....

Solve multiplication problems, including scaling and correspondence problems

1 Kyle's dad is making Kyle a tree house. Firstly, he makes a model of the tree house which is **50 cm** high. When the tree house is made, it is **eight** times as high as the model. How high is the real tree house?

cm

2 Coco The Clown has **14** different red noses and **4** different wigs. How many different clown outfits can he wear?

3 Grumpy the Giant is **20 metres** tall. Jack is **ten** times smaller than Grumpy. How tall is Jack?

m

4 Jessica made a huge ice cube that weighed **720 grams**. In the 3 hours after it was taken out of the freezer, it shrank by **8** times its original size. How heavy was the ice cube after 3 hours?

g

5 Mr Potato Head has **13** different noses and **8** different mouths. How many different Mr Potato Heads can be made?

Name ....

It may be appropriate for children to use exercise books or paper to record their answers, working out or explanations.

MASTERING

Multiplication and division

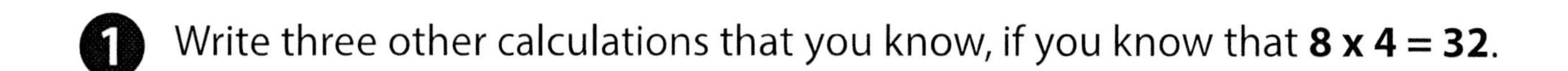

1 Write three other calculations that you know, if you know that **8 x 4 = 32**.

☐ × ☐ = ☐ ☐ ÷ ☐ = ☐

☐ ÷ ☐ = ☐

2 Find the missing digits in this multiplication square.

×			5	2	3
4	32	16			
			15		
5				10	
				4	
					24

3 Jessica buys **3** chocolate bars.
She pays using a **£5** note and gets **£2.90** change.

How much does each chocolate bar cost?

☐ p

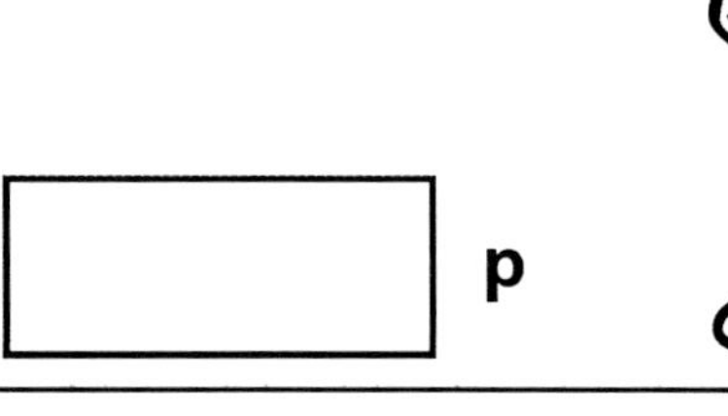

Name ....

4 Use what you know about **5 × 6** and **10 × 6** to help you find **15 × 6**.
Show your working out.

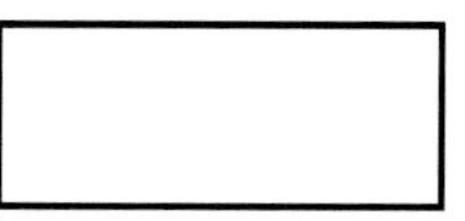

5 Sarah has **24** photos.
She wants to put them all into a rectangular frame. Find the different ways she can organise her photos so that they are in a rectangle
One has been done for you.

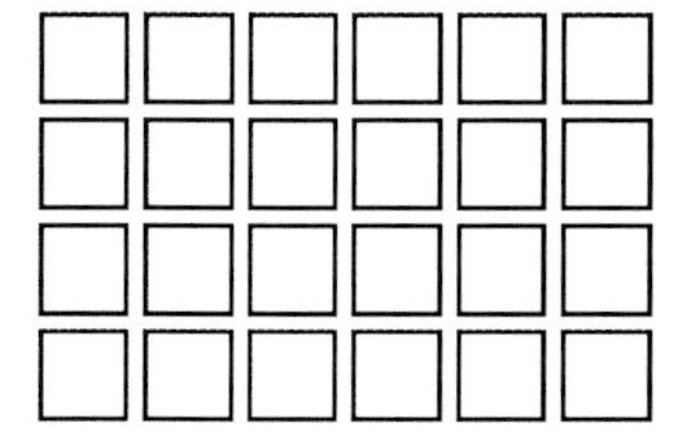

6 Kyle has finished his maths work.
Use the inverse operation to help him check his calculation.
Show the inverse operation you have used and correct any mistakes he has made.

24 × 3 = 76

36 ÷ 4 = 12

86 × 8 = 688

375 ÷ 5 = 76

Name ....

7 Jessica takes a packed lunch to school every day.
She always has a banana in her packed lunch.
A banana costs **12p**.

a How much does Jessica's mum spend on bananas for five packed lunches?

[ ] p

b In the Spring Term, there are **12** weeks.

How much will Jessica's mum spend on bananas in the Spring Term?

£ [ ]

c Use your answer to **a** to help you work out how much Jessica's mum spends on bananas in a year, if Jessica goes to school for **39** weeks.

£ [ ]

---

8 Sarah is sorting some pencils into **8** pencil pots.
She has between **20** and **30** pencils.

When she puts **2** into each pot she has **8** left over.

How many pencils does Sarah have?

[ ]

How many does she need to put into each pot so that she has none left over?

[ ]

Name ....

9 Mrs Smith, Mrs Patel and Mr Stanton each hear a group of children reading every week. Mrs Smith's group is the smallest. Mrs Patel's group is the largest.

a Mrs Smith buys **23** biscuits.
She shares them between the **7** children in her reading group.

How many biscuits will each child get?

How many biscuits are left over?

b Mrs Patel has **28** chocolate biscuits.
She shares them between the children in her reading group.
Each child gets **3** biscuits and there is **one** biscuit left over.

How many children are there in Mrs Patel's group?

c Mr Stanton shares some biscuits with the children in his group.

Each child receives **4** biscuits and there are **3** biscuits left over.

How many biscuits did Mr Stanton start with?

Name ....

10 The ×, ÷ and = signs are missing from these equations. Fill in the missing signs.

5 ☐ 8 ☐ 2 ☐ 20

2 ☐ 3 ☐ 5 ☐ 30

3 ☐ 24 ☐ 8

---

11 Use =, < or > in the boxes below.

8×3 ☐ 4×6

5×3 ☐ 4×4

$40 \div 4$ ☐ $35 \div 5$

2×4 ☐ $36 \div 4$

3×4 ☐ 2×6

Name ....

12 Kyle says that **8** is an even number and all the multiples of **8** are also even. This means that as **3** is an odd number, all the multiples of **3** are also odd.

Is Kyle correct? **Yes / No**

Use examples to explain your answer.

....

....

13 Fill in the boxes below using **8** different whole numbers.

□ × □ = **30**

□ × □ = **30**

□ × □ = **30**

□ × □ = **30**

14 Find the missing digits in these calculations.

a

$$\begin{array}{r} 2\,\square \\ \times \quad 5 \\ \hline 1\,3\,5 \\ \hline \end{array}$$

b

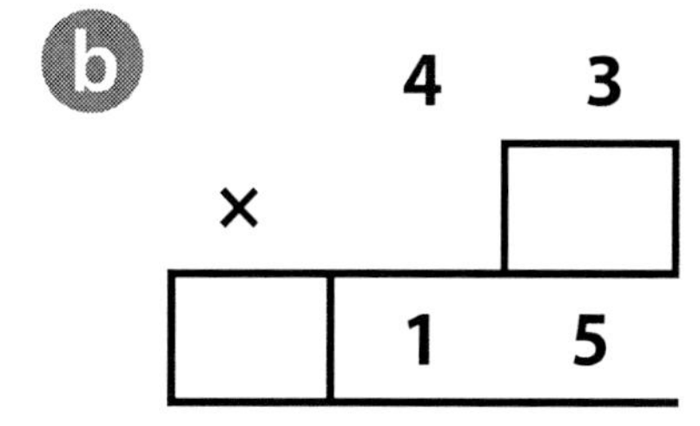

$$\begin{array}{r} 4\,3 \\ \times \quad \square \\ \hline \square\,1\,5 \\ \hline \end{array}$$

c

$$\begin{array}{r} \square\,5 \\ \times \quad 4 \\ \hline \square\,4\,0 \\ \hline \end{array}$$

d

$$\begin{array}{r} 2\,\square\,6 \\ \times \quad \square \\ \hline 5\,5\,2 \\ \hline \end{array}$$

Name ....

15 Fill in the missing numbers to make these number sentences correct.

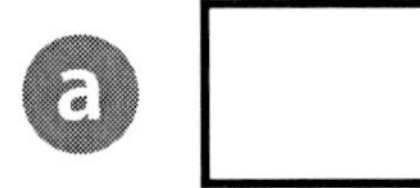

a) $\square \times 4 = 32$

d) $\square \div 5 = 8$

b) $3 \times \square = 18$

e) $72 \div 8 = \square$

c) $5 \times 9 = \square$

f) $5 \times \square = 5$

---

16 a) Use the numbers **2**, **4** and **5**.
Make as many calculations as you can by putting the numbers in these boxes.

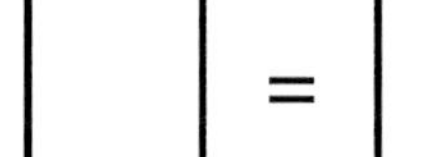
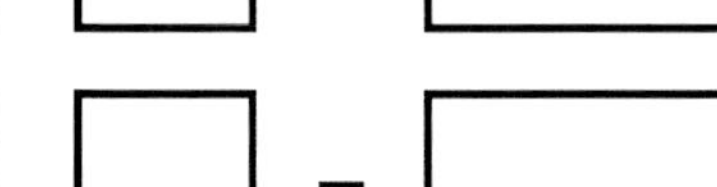
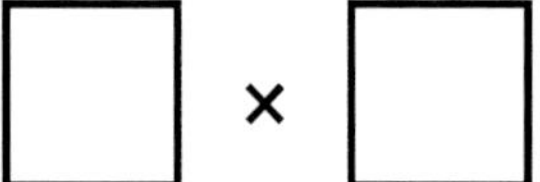
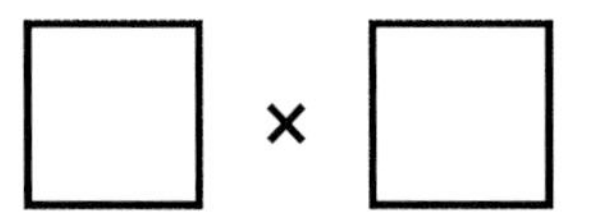
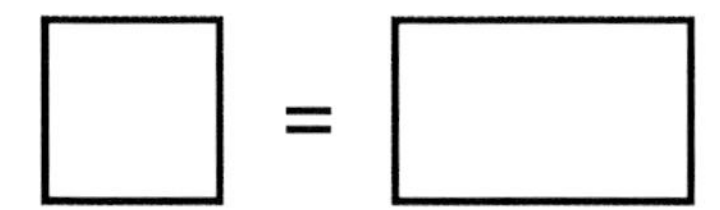
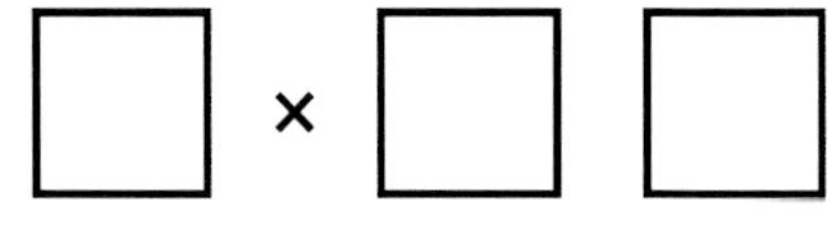

$\square \times \square\square = \square$

$\square \times \square\square = \square$

$\square \times \square\square = \square$

$\square \times \square\square = \square$

$\square \times \square\square = \square$

$\square \times \square\square = \square$

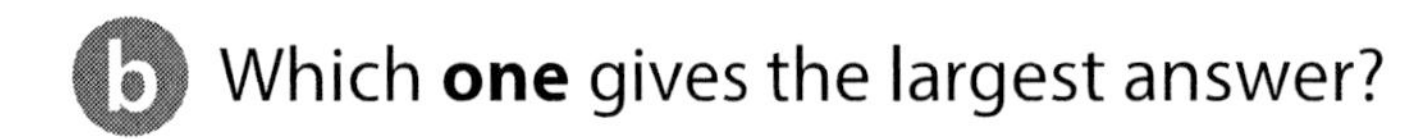

b) Which **one** gives the largest answer?

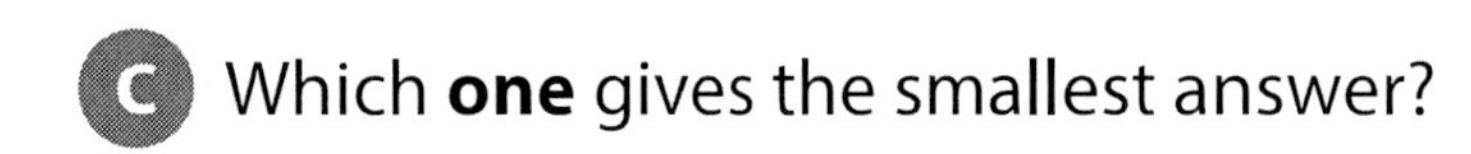

c) Which **one** gives the smallest answer?

Name ....

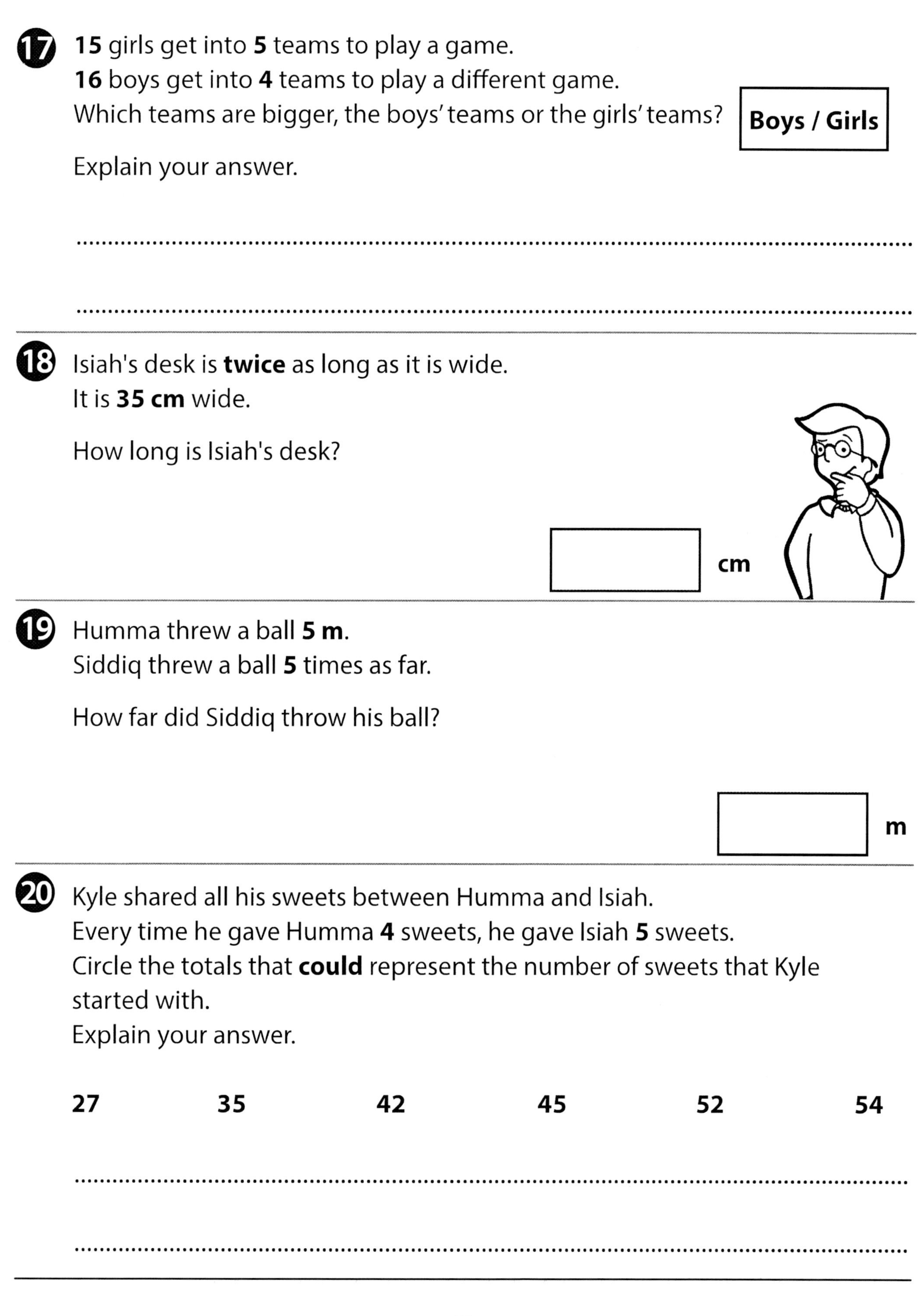

17 **15** girls get into **5** teams to play a game.
16 boys get into **4** teams to play a different game.
Which teams are bigger, the boys' teams or the girls' teams? **Boys / Girls**

Explain your answer.

....

....

18 Isiah's desk is **twice** as long as it is wide.
It is **35 cm** wide.

How long is Isiah's desk?

[ ] **cm**

19 Humma threw a ball **5 m**.
Siddiq threw a ball **5** times as far.

How far did Siddiq throw his ball?

[ ] **m**

20 Kyle shared all his sweets between Humma and Isiah.
Every time he gave Humma **4** sweets, he gave Isiah **5** sweets.
Circle the totals that **could** represent the number of sweets that Kyle started with.
Explain your answer.

27 **35** **42** **45** **52** **54**

....

....

Name ....

21 Siddiq says that every alternate multiple in the **4** times table is also in the **8** times table.
Is Siddiq correct? **Yes / No**

Use examples to explain your answer.

....

....

22 The tables in Mrs Higgins' classroom have **3** legs and the chairs have **4** legs.
Mrs Higgins counted the legs in her classroom.

There were **152** legs.

There were **4** times as many chairs as tables.

How many chairs and tables are there in Mrs Higgins' classroom?

chairs [ ] **tables** [ ]

23 There are **eight** plums in a punnet.
How many punnets of plums do I need to buy so that all **30** children in my class can have **one** each?

[ ]

  Name ....

Sarah says: **23 × 2 = 46**

Humma says: **31 × 3 = 93**

The girls decide that if you multiply a **2-digit** number by a **1-digit** number the answer will always be **2-digits**.

Are the girls correct? **Yes / No**

Use examples to explain your answer.

....

....

---

25 I know that **8 × 4 = 32**. So, what is:

80 × 4 

8 × 40 

8 × 20 

8 × 24 

Kyle is building a wall out of Lego.
He puts **8** bricks on the bottom layer. He has **56** bricks altogether.
How many layers of bricks can Kyle make if he uses all of his bricks?

Name ....

27 Siddiq wants to make **160** using the digits **2**, **3** and **5**.

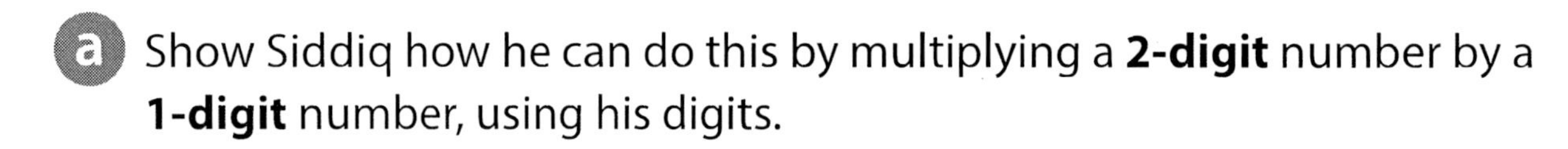

a Show Siddiq how he can do this by multiplying a **2-digit** number by a **1-digit** number, using his digits.

☐ ☐ × ☐ = **160**

b Siddiq uses his digits to make **two** odd numbers.
Show Siddiq how to make **two** odd numbers by multiplying a **2-digit** number by a **1-digit** number, using his digits.

☐ ☐ × ☐ = ☐

☐ ☐ × ☐ = ☐

---

28 Add half of **40** to double **8**.

☐

29 Humma wants to make a rectangular array out of hoops in the playground.
She has **48** hoops.
Find **4** different ways of arranging the hoops into a rectangle.

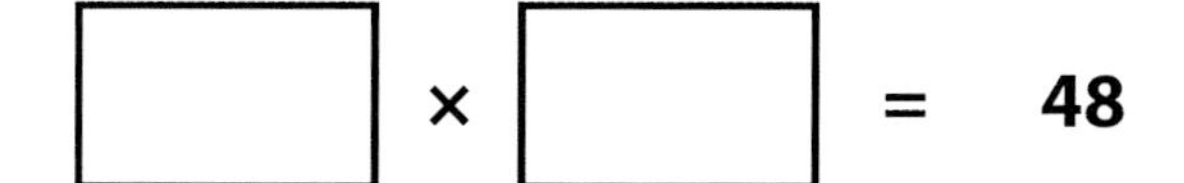

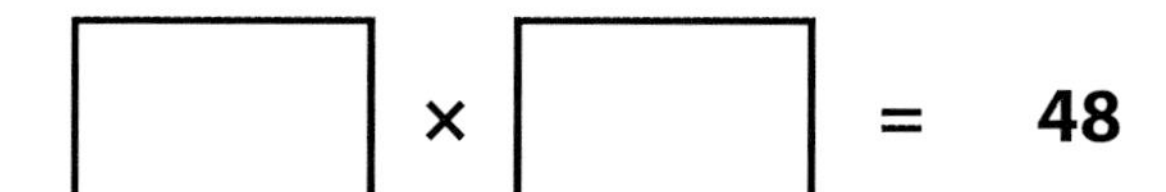

30 Will the size of these answers be larger or smaller than **100**?
Circle the right answer for each.

a	**25 × 5**	**larger**	**smaller**
b	**30 × 4**	**larger**	**smaller**
c	**400 ÷ 2**	**larger**	**smaller**
d	**8 × 10**	**larger**	**smaller**

Name ....

31 Subtract double **20** from half of **100**.

32 Multiply half of **64** by double **4**.

33 Divide double **8** by half of **8**.

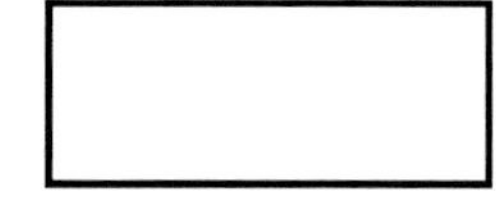

34 Write numbers in the boxes to make this calculation correct.

×		4
8	160	

Name ....

It may be appropriate for children to use exercise books or paper to record their answers, working out or explanations.

NUMBER

Fractions

Recognise that tenths arise from dividing an object into 10 equal parts

1. Humma eats $\frac{1}{10}$ of her pizza. How many **tenths** are left?

2. Kyle has **10** cars. He gives **2** to his brother. What fraction of cars does Kyle have left?

3. Sarah has **10** DVDs. She watches **3** with her mum. Write the number of DVDs they have watched as a fraction.

4. Jessica's birthday cake is cut into **10** pieces. If **6** children at her party have a piece, what fraction of the cake is left?

5. The joiner cuts a piece of wood into **10** pieces. He uses **7** pieces of wood to make a chair. Write how many pieces are left as a fraction.

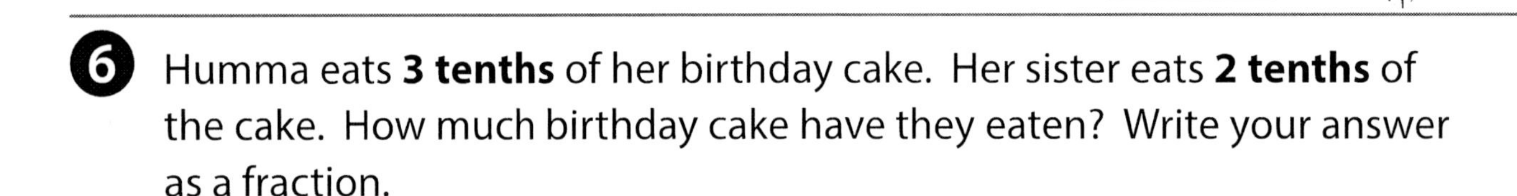

6. Humma eats **3 tenths** of her birthday cake. Her sister eats **2 tenths** of the cake. How much birthday cake have they eaten? Write your answer as a fraction.

Name ....

Recognise, find and write unit fractions of a discrete set of objects

1 **One half** of an orange has **8** segments. How many segments would there be in a whole orange?

2 There are **12** sheep in a field. **One quarter** of the sheep are eating grass. How many sheep are not eating grass?

3 Sarah has **15** teddies. She gives $\frac{1}{3}$ of her teddies to her little sister. How many teddies has she given to her sister?

4 Humma has **30** books. She puts her books in piles. Each pile is a **sixth** of the total amount. How many books are in each pile?

5 Kyle's bar of chocolate had **24** pieces. He wanted to break the bar into **eighths**. How many pieces would be in each **eighth**?

6 A Premier League football costs **£18** in the sale. It is $\frac{1}{3}$ of the original price. How much was the football originally?

£

Name ....

Recognise, find and write non-unit fractions of a discrete set of objects

1 Humma has **15** dresses in her wardrobe. $\frac{2}{3}$ of her dresses are pink. How many of Humma's dresses are pink?

2 Kyle had **40** football cards. He wanted to find out how many cards would make up **three quarters** of his collection. Explain how Kyle would work this out. What would his answer be?

3 Mrs Dewhurst bakes **27** miniature apple pies. Her family eats $\frac{2}{3}$ of them. How many pies are left?

4 Sarah has **30** fizzy sweets. She eats **three fifths** of her fizzy sweets. How many sweets has she eaten?

5 Siddiq had **54** DVDs. He sold **five sixths** of his collection. How many DVDs did he sell?

Name ....

Understand equivalence in unit and non-unit fractions

1 The twins were sharing a cake. One had **half** of the cake and the other had **two quarters**. Did they have the same amount?

Yes / No

Explain your answer.

....

....

2 Isiah cuts a cake into **6** equal slices. He eats **2** slices of his cake. Has Isiah eaten $\frac{1}{2}$ of his cake?

Yes / No

Explain your answer.

....

....

3 Kyle ordered a pizza. The pizza chef asked if he would like the pizza cutting into **quarters** or **eighths**. Kyle said, "Just into **quarters**, because I'm hungry enough to eat **4** pieces, but not hungry enough to eat **8** pieces."

Explain why Kyle's answer didn't make sense.

....

....

4 Siddiq has saved **£6**. He says, "If I spend **two thirds** of my money, that's the same as spending **three sixths** of my money." Is he correct?

Yes / No

Explain your answer.

....

....

Name ....

Understand the relation between unit fractions as operators and division by integers

1 Jessica had **20** sweets. She wanted to share the sweets between herself and **3** friends by finding a **quarter** of the sweets. How many would each person have?

---

2 Kyle has **30** iced cakes. He gives them out equally to himself and **5** friends by working out $\frac{1}{6}$ of the cakes. How many would each person have?

---

3 Mr Redmond has **24** pots of paint. He shares them between **8** children. He does this by finding **one eighth** of the paint pots. How many pots of paint would each child have?

---

4 Siddiq works out that **70 ÷ 5 = 14**. What is **one fifth** of **70**?

---

5 Mrs Jameson asks her class to work out a **sixth** of **42** and a **fifth** of **60** and then add the **two** answers together. Explain using words or numbers how you would do this.
Write the answer too.

....

....

Name ....

Add fractions with the same denominator within one whole

1 Mrs Holden asks her class what $\frac{1}{3}$ add $\frac{1}{3}$ would be. What answer should they give?

$\frac{\square}{\square}$

---

2 Isiah drinks $\frac{1}{4}$ of his bottle of juice with his lunch. He then drinks another $\frac{2}{4}$ of his juice. How much of his juice has Isiah drunk altogether? Give your answer as a fraction.

$\frac{\square}{\square}$

---

3 Kyle ate **two fifths** of his chocolate bar at lunchtime. He ate another $\frac{2}{5}$ on the way home. What fraction of his chocolate bar did he eat altogether?

$\frac{\square}{\square}$

---

4 Humma runs $\frac{3}{7}$ of her way home from school. She then walks $\frac{2}{7}$ more. What fraction of the distance has she travelled altogether?

$\frac{\square}{\square}$

---

5 **Two eighths** of Farmer Fred's cows are moved to field A. Farmer Fred then moves another $\frac{5}{8}$ of his cows to field A. Write the fraction of cows that are in field A now.

$\frac{\square}{\square}$

Name ....

Subtract fractions with the same denominator within one whole

1 Jessica needs to subtract $\frac{1}{3}$ from $\frac{2}{3}$. What would her answer be?

$\frac{\square}{\square}$

2 $\frac{4}{5}$ of the herd of elephants are cooling down in the water. $\frac{2}{5}$ of the elephants get out of the water to find food. What fraction of the elephants are left in the water?

$\frac{\square}{\square}$

3 **Three quarters** of Year 3 are playing outside. $\frac{1}{4}$ of Year 3 go back inside to their classroom. What fraction of Year 3 are still outside?

$\frac{\square}{\square}$

4 $\frac{5}{6}$ of Siddiq's cupcakes are iced with either blue or red icing. $\frac{2}{6}$ of his cakes have red icing. What fraction of his cupcakes have blue icing?

$\frac{\square}{\square}$

5 Isiah eats $\frac{5}{8}$ of his chocolate bar and gives **two eighths** of his chocolate bar to his brother. How much chocolate does he have left?

$\frac{\square}{\square}$

Name ....

Add and subtract fractions with the same denominator within one whole

1 Kyle adds $\frac{1}{4}$ and $\frac{2}{4}$. What is his answer?

$\frac{\quad}{\quad}$

2 Sarah took $\frac{1}{3}$ of the pencils from the box. Siddiq took another **third** of the pencils from the box. What fraction of the pencils did Sarah and Siddiq take out altogether?

$\frac{\quad}{\quad}$

3 Jessica is trying to work out what $\frac{2}{6}$ add **three sixths** would be. What fraction should she write?

$\frac{\quad}{\quad}$

4 Mr Jenson takes $\frac{4}{5}$ of the school bean bags from the bean bag box. He gives out **three fifths** to his class. What fraction of bean bags does he have left?

$\frac{\quad}{\quad}$

5 Kyle walks his dog, Deefer, $\frac{2}{5}$ of the way around the park. They have a rest and then walk another $\frac{2}{5}$ of the way around the park. What fraction do they still need to walk to go the whole way around the park?

$\frac{\quad}{\quad}$

6 Siddiq spends $\frac{3}{8}$ of his money on a t-shirt. He then buys a magazine with $\frac{1}{8}$ of his money. His grandma gives him $\frac{2}{8}$ of his money back. How much money does he have now? Write your answer as a fraction.

$\frac{\quad}{\quad}$

Name ....

Compare and order unit fractions and non–unit fractions with the same denominator

1 Mr Badal asks his class which fraction is bigger: $\frac{3}{4}$ or $\frac{2}{4}$. Which one should they choose?

$\frac{\square}{\square}$

---

2 The cat ate **two thirds** of her food. The dog ate $\frac{1}{3}$ of his food. Which pet ate the larger fraction of their food?

---

3 Jessica had **one fifth** of the bottle of cola and Isiah had **two fifths** of the bottle. Who had more?
Explain your answer.

Jessica / Isiah

....

....

---

4 Kyle had $\frac{2}{6}$ of the pizza, Siddiq had $\frac{1}{6}$ and Humma had $\frac{3}{6}$. Who had the most and who had the least?

most = ☐ **least =** ☐

---

5 Put these fractions in order of size, starting with the smallest:

$\frac{3}{8}$ $\frac{4}{8}$ $\frac{1}{8}$ $\frac{7}{8}$

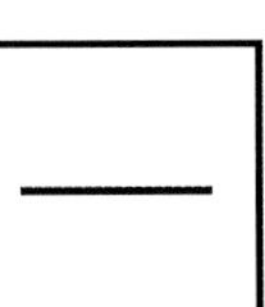

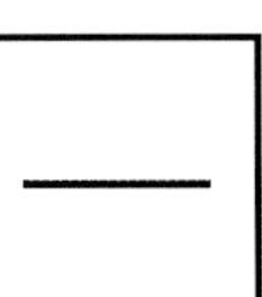

$\frac{\square}{\square}$

Name ....

Solve problems involving fractions

1 Sarah has a **200 ml** glass of milk. She drinks **half** of it. How many **millilitres** of milk does she have left?

ml

---

2 Kyle wants to find $\frac{1}{4}$ of **£20**. What would his answer be in **pounds**?

£ 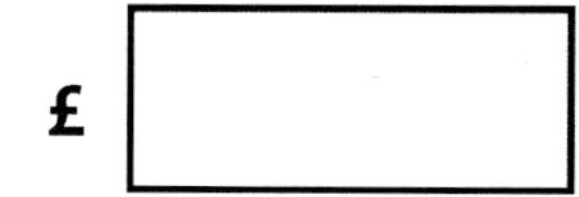

---

3 Humma has **25** sweets. **One fifth** of them are chewy. How many chewy sweets does she have?

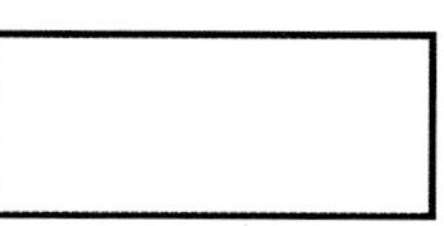

---

4 There are **30** children in a class. $\frac{2}{3}$ of the class are girls. How many boys are there in the class?

---

5 A shop had **12 kilograms** of apples. It sold $\frac{5}{6}$ of the apples on Tuesday. How many **kilograms** of apples did the shop have left?

---

6 Circle the larger amount below.

$\frac{1}{2}$ **of £24** or $\frac{1}{4}$ **of £36**

Name ....

Solve problems involving fractions

1 Kyle ate $\frac{4}{8}$ of his pizza on Monday night. On Tuesday, he ate another $\frac{3}{8}$. What fraction of his pizza did he eat on Monday and Tuesday altogether?

2 Mrs Bunter cut her cake into **6** equal pieces. She gave her daughter **2** pieces. What fraction of the cake did her daughter have?

3 Siddiq has **one half** of the play dough. Isiah has **two quarters**. Who has the most play dough? Explain your answer.

....

....

4 If **60 ÷ 5 = 12**, what is **one twelfth** of **60**?

5 Humma says, "A **quarter** of the numbers on a **hundred** square are even." Is she correct and how do you know?

....

....

6 In the shop *Look Beautiful*, a dress has been reduced by **one eighth**. It was **£32**. How much is it now?

£

Name ....

Solve problems involving fractions

1 How many **tenths** are shaded.

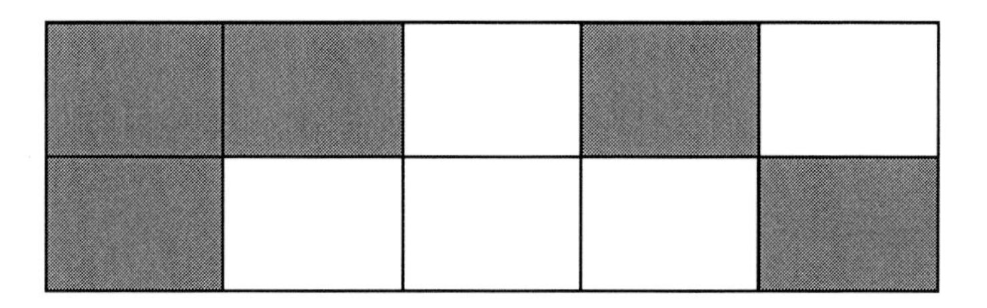

2 Complete the shading on the shape so that **eight tenths** of the shape are shaded.

3 Humma has worked out that **4** divided by **10** is **0.4**. What would the answer be to **5** divided by **10**?

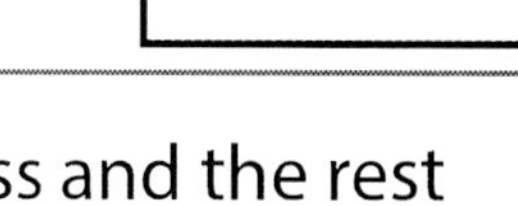

4 There are **8** horses in a field. **5** of the horses are eating grass and the rest are galloping around the field. What fraction of the horses are galloping?

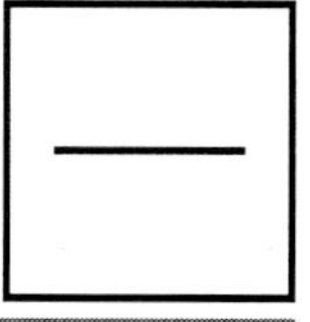

5 Mrs Williams asks her class to choose a fraction that is the same as $\frac{1}{8}$. Can you name **2** fractions that they could choose?

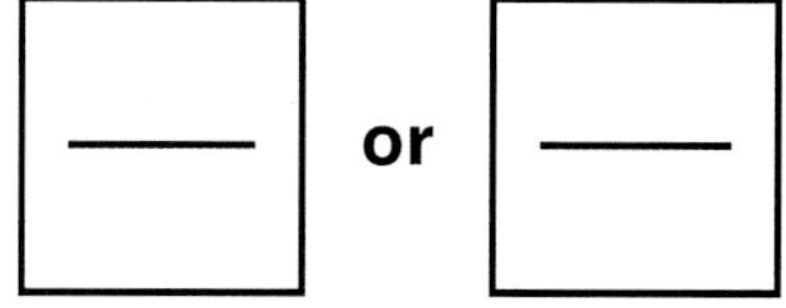

6 A pair of trainers has a **quarter** off the price in the sale. If the trainers cost **£48** before the sale, how much are they in the sale?

£

Name ....

It may be appropriate for children to use exercise books or paper to record their answers, working out or explanations.

MASTERING

Fractions

1 Sarah had **24** marbles.

She gave $\frac{1}{4}$ of the marbles to Kyle and $\frac{1}{6}$ of what was left to Siddiq.

She kept the rest for herself. How many marbles did each child have?

Sarah ☐ **Kyle** ☐ **Siddiq** 

---

2 Mark the position of these fractions on the number lines below.

a $\frac{1}{4}$ $\frac{2}{4}$ $\frac{3}{4}$

0 ———————————————— **1**

b $\frac{1}{6}$ $\frac{1}{3}$ $\frac{1}{2}$

0 ———————————————— **1**

c How big is the interval from:

$\frac{1}{6}$ **to** $\frac{1}{3}$ ☐

$\frac{1}{6}$ **to** $\frac{1}{2}$ ☐

Name ....

3 Isiah looked at his watch. It was **3 o'clock**.
He noticed that the hands showed exactly a **quarter** of the clock face.

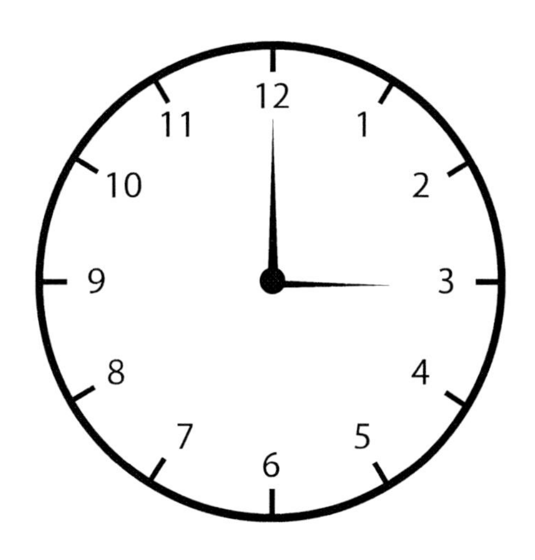

What is the only other time that the hands would show exactly a quarter of the clock face?

---

4 Colour the fraction shown for each of these shapes.

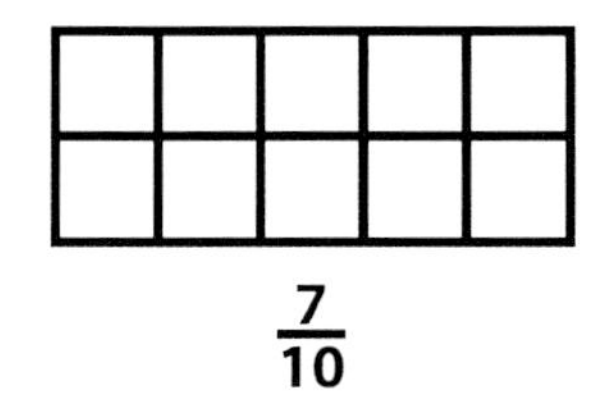

$\frac{7}{10}$

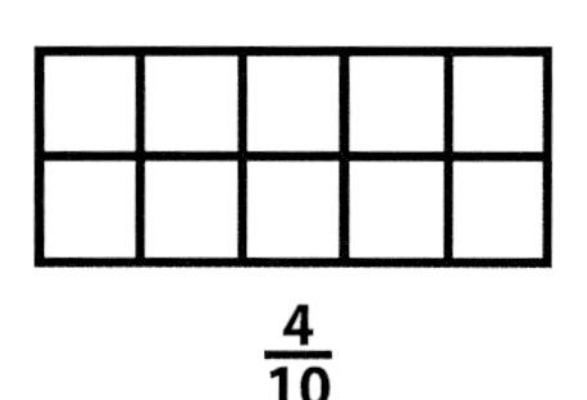

$\frac{4}{10}$

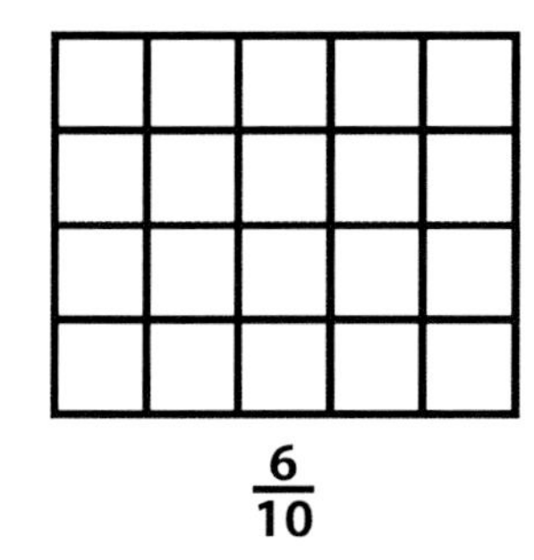

$\frac{6}{10}$

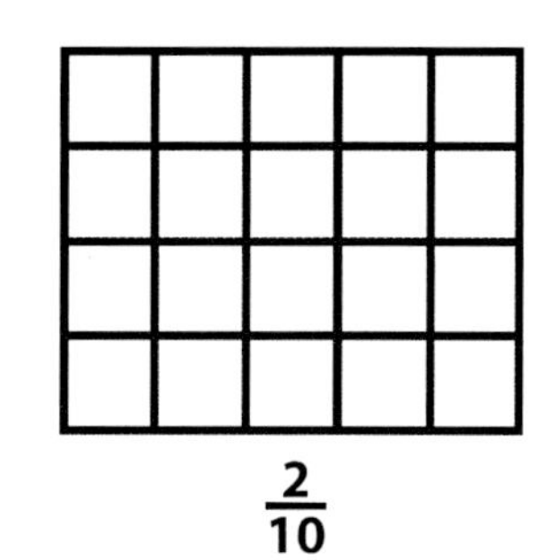

$\frac{2}{10}$

Name ....

5 Kyle has divided these shapes into the fractions shown.
He has made some mistakes.
If Kyle is not correct, explain his mistake.

	Shape	Fraction	Correct?	Reason it is not correct
a		$\frac{1}{2}$	Yes / No	
b		$\frac{1}{3}$	Yes / No	
c		$\frac{1}{2}$	Yes / No	
d		$\frac{1}{5}$	Yes / No	

Name ....

6 Siddiq had **two** lengths of pipe. He hid them behind a wall and asked Isiah to guess which was the longer one.

You can see part of each pipe here.

a Which length of pipe has the longer total length? **Pipe 1 / Pipe 2**

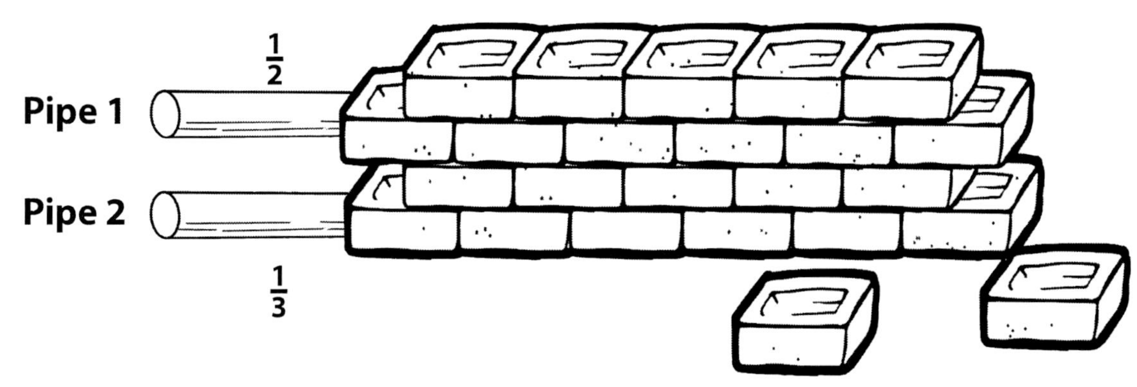

b How do you know?

....

....

7 Humma ate $\frac{1}{5}$ of her sweets. She had **8** sweets left.
How many sweets did she have to begin with?

8 Put these fractions in order of size starting with the smallest.

$\frac{2}{5}$ $\frac{1}{10}$ $\frac{3}{10}$

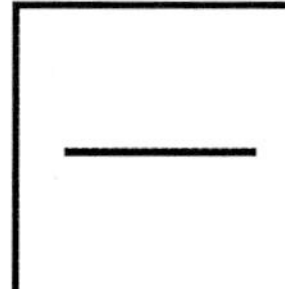
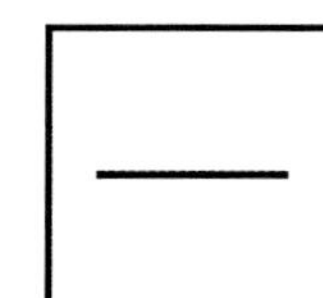

9 Isiah has this set of marbles.
R = red. B = blue. G = green.

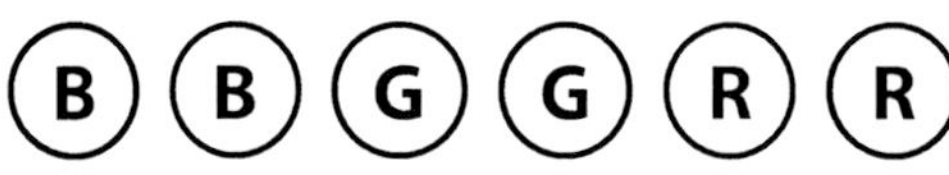

a Find the fraction of each colour of marble.

red —— **blue** —— **green** 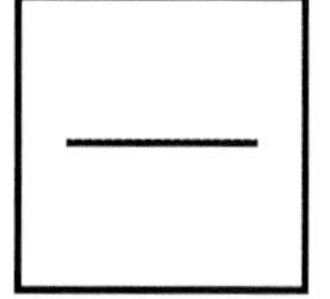

b Isiah says, "If I take away **4** red marbles, I will have to change all the fractions for the other colours of marbles that are left."

Is Isiah correct? **Yes / No**

c Write the fraction for each colour of marbles now.

red —— **blue** 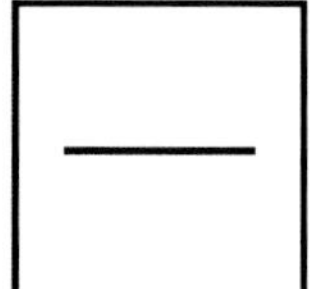 **green** ——

10 Complete the number sentence correctly in 4 different ways.

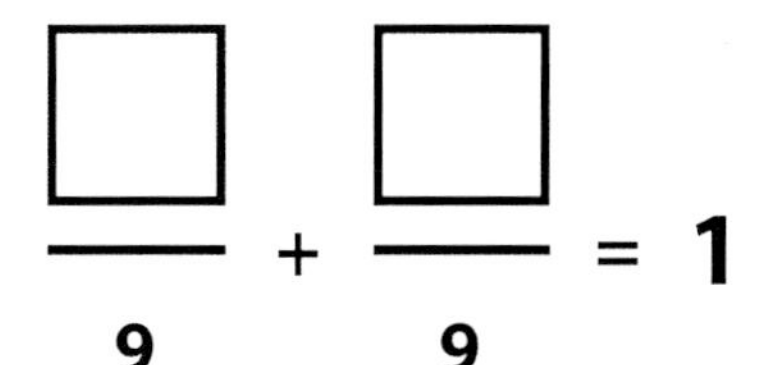

$\frac{\square}{9} + \frac{\square}{9} = 1$ $\frac{\square}{9} + \frac{\square}{9} = 1$

$\frac{\square}{9} + \frac{\square}{9} = 1$ $\frac{\square}{9} + \frac{\square}{9} = 1$

11 Put a circle around the fraction which is equal to $\frac{3}{4}$.

$\frac{16}{20}$ $\frac{65}{100}$ $\frac{2}{3}$ $\frac{75}{100}$ $\frac{7}{10}$

12 Sarah says that the diagrams below show that $\frac{1}{6} > \frac{1}{3}$.
Is she correct?

Yes / No

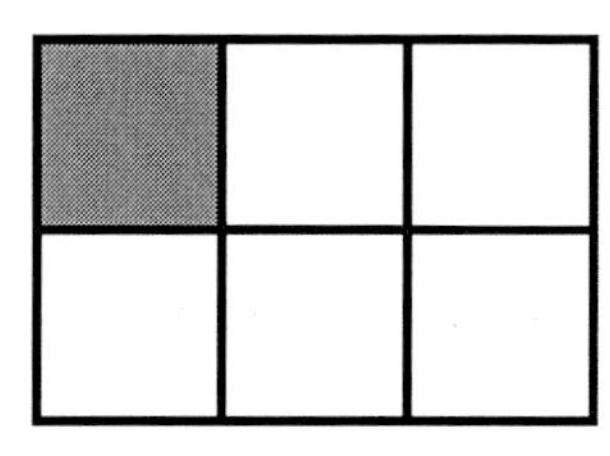

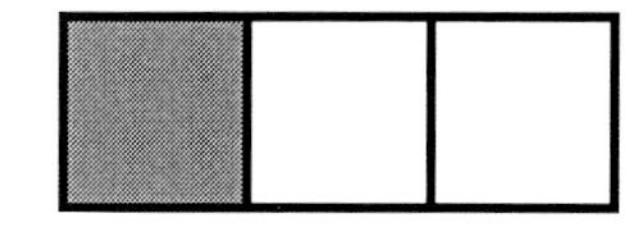

Explain your answer.

....

....

Name ....

13 Siddiq has done some maths homework.
He has made some mistakes.
Mark his work, (× or ✓) and correct any mistakes he has made.

a $\frac{3}{10}$ **of 300 g = 30 g** ☐

correction ....

b $\frac{2}{5}$ **of 400 ml = 150 ml** ☐

correction ....

c $\frac{7}{10}$ **of 500 m = 350 m** ☐

correction ....

14 Jessica gives **12** sweets to Isiah. She says, "I have given you $\frac{2}{3}$ of my sweets."
How many sweets did Jessica have to start with?

15 **a** What fraction of the shape is shaded?

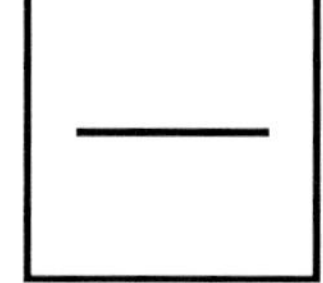

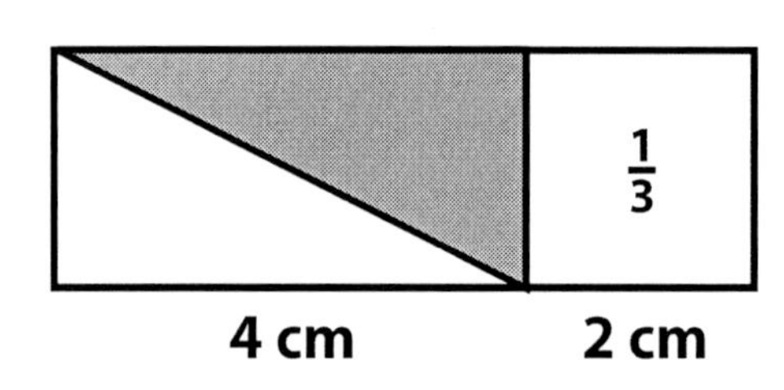

b Explain your answer.

....

....

Name ....

 Give **three** examples of fractions that are smaller than a **half**.

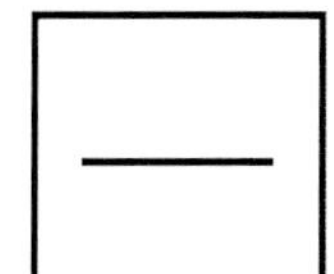 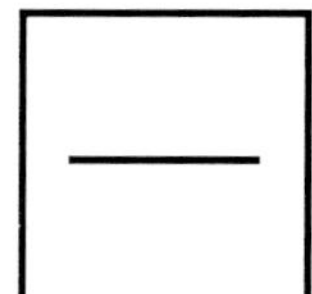 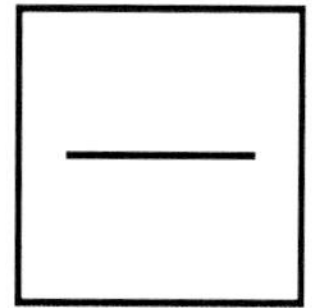

Shade in squares to show how you know this.

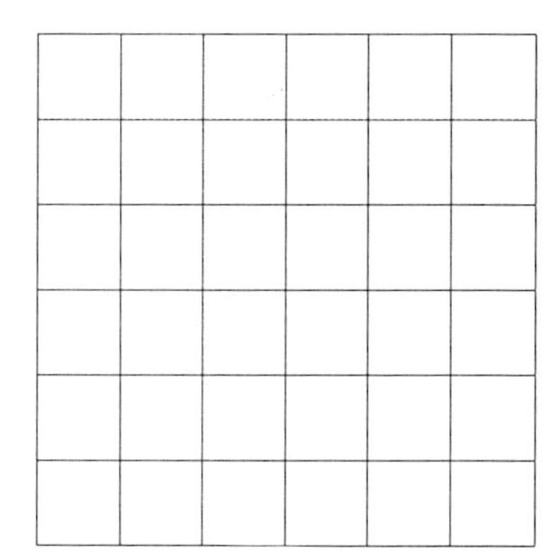 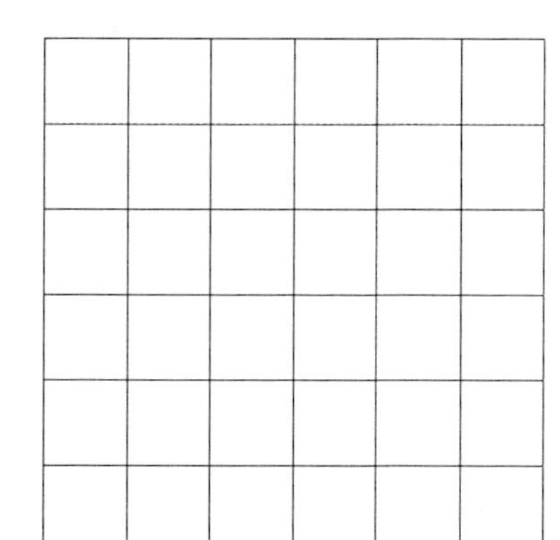 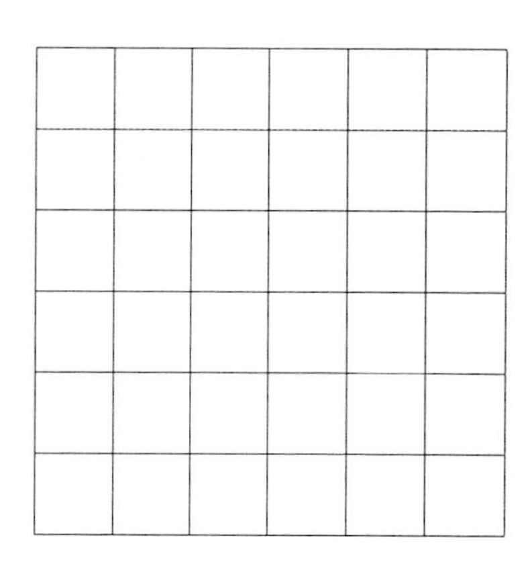

---

17 Complete these number sentences to make them correct.

a $\frac{1}{10} + \frac{\square}{10} = 1$

b $\frac{3}{10} + \frac{7}{\square} = 1$

c $\frac{\square}{10} + \frac{8}{10} = 1$

d $\frac{\square}{10} + \frac{\square}{10} = 1$

18 Kyle is very generous. He was given **24** sweets for his birthday.
On Monday, he gave $\frac{1}{8}$ of his sweets to his mum and ate only **one** himself.

On Tuesday, he gave $\frac{1}{5}$ of his remaining sweets to his dad and ate only **one** himself.

On Wednesday, he gave $\frac{1}{3}$ of his remaining sweets to his sister and ate only **one** himself.

On Thursday, he gave $\frac{1}{3}$ of his remaining sweets to his best friend and ate the rest himself.

a How many sweets did Kyle give away each day?
Complete the table below.

Monday	Tuesday	Wednesday	Thursday

b How many sweets did Kyle eat on Thursday?

c What fraction of the 24 sweets did Kyle eat on Thursday?

19 Sarah cut a cake up like this:

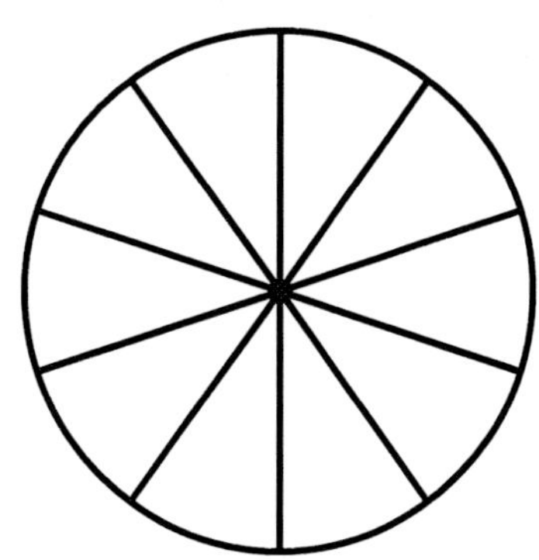

a What fraction of the cake does each section represent?

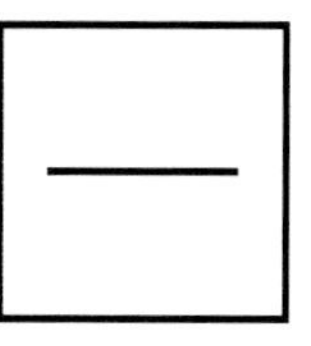

b Divide these cakes up into eighths and sixths.

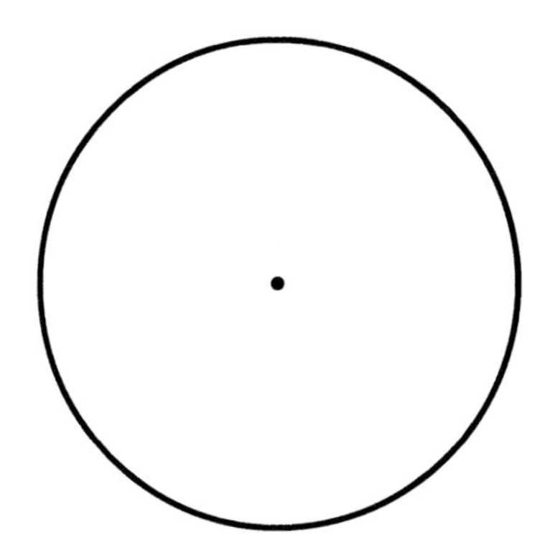

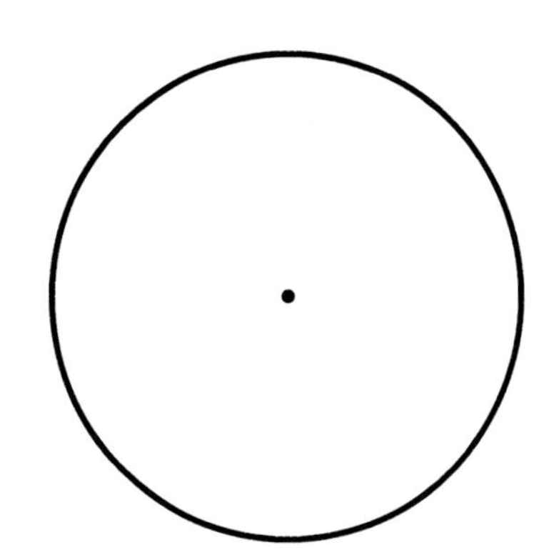

c Complete this number sentence so that it is correct using the symbols > and <.

$\frac{3}{10}$ 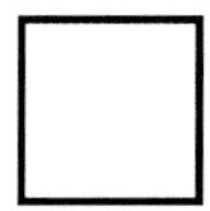 $\frac{3}{6}$ 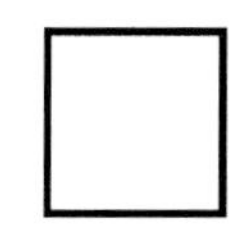 $\frac{1}{8}$

Name ....

20 Complete these sequences.

 $\frac{3}{10}$ $\frac{4}{10}$ $\frac{5}{10}$ 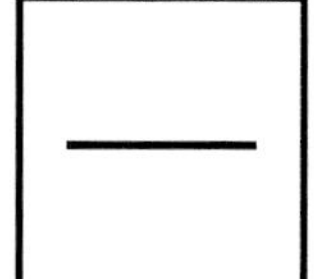

b $\frac{15}{10}$ $\frac{14}{10}$ $\frac{13}{10}$ 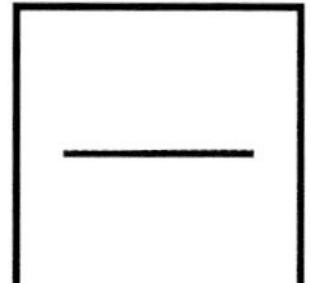

---

21 Write a number in each box to make the number sentence correct.

 $\frac{1}{4}$ of **24** = $\frac{1}{2}$ of 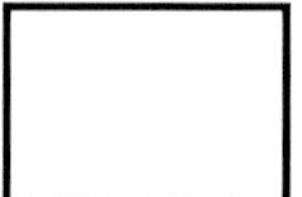

b $\frac{1}{3}$ of ☐ = $\frac{1}{2}$ of **4**

c $\frac{1}{10}$ of **100** = ☐ of **20**

d 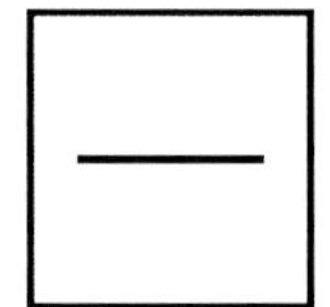 of **25** = $\frac{1}{10}$ of **50**

---

22 Order these fractions from smallest to largest.

$\frac{1}{5}$ $\frac{1}{3}$ $\frac{1}{10}$ $\frac{1}{2}$ $\frac{1}{4}$ $\frac{1}{7}$

☐ ☐ 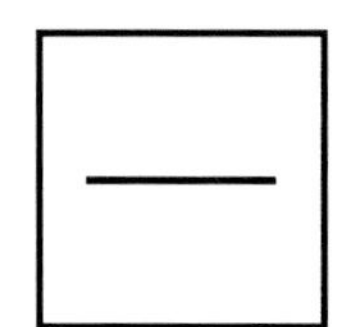 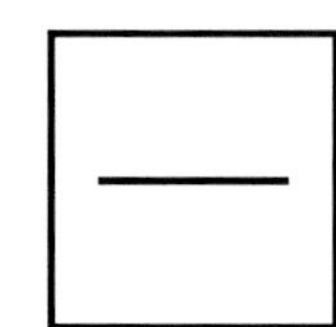 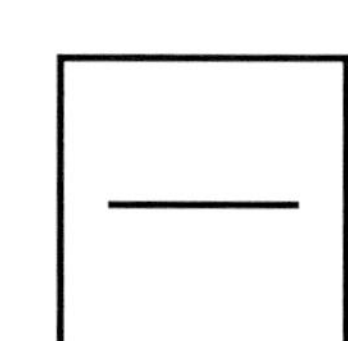 ☐

smallest **largest**

Name ....

23 Write a subtraction sentence that you can find if you know $\frac{2}{10} + \frac{3}{10} = \frac{5}{10}$.

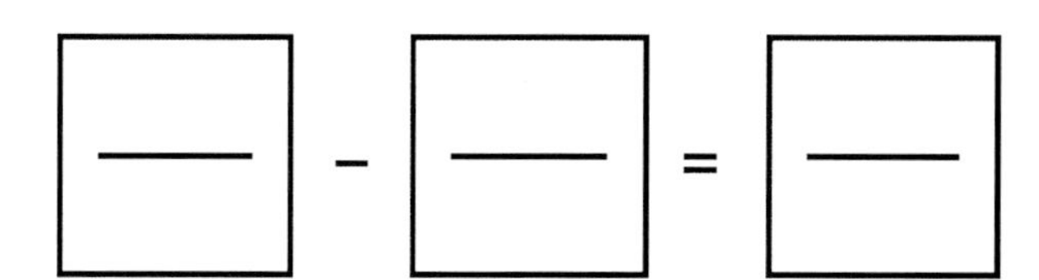

24 Find $\frac{2}{8}$ of **16** and $\frac{1}{4}$ of **16**.

What do you notice?

....

Write down 3 other pairs of fractions that are equal.

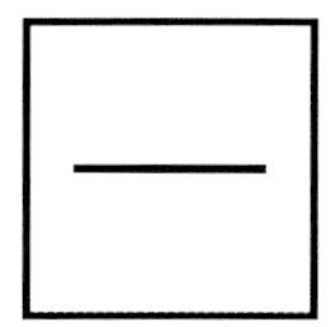 **and** 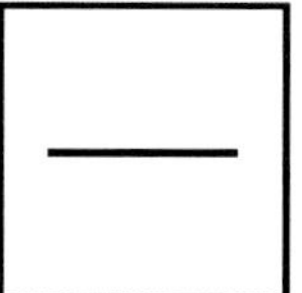   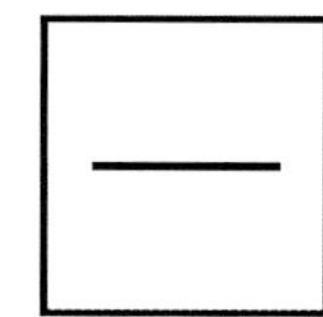 **and** 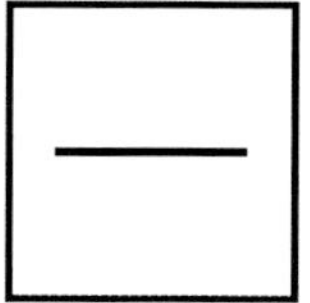

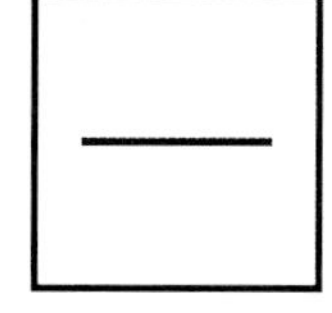 **and** 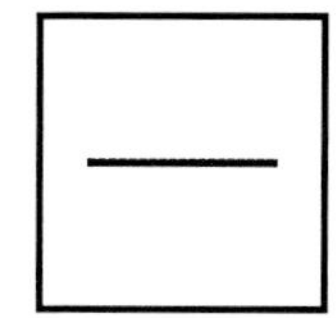

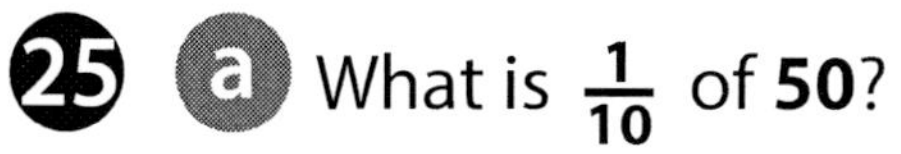
25 a What is $\frac{1}{10}$ of **50**?

b Use your answer to find:

$\frac{2}{10}$ of **50** 

$\frac{3}{10}$ of **50** 

Name ....

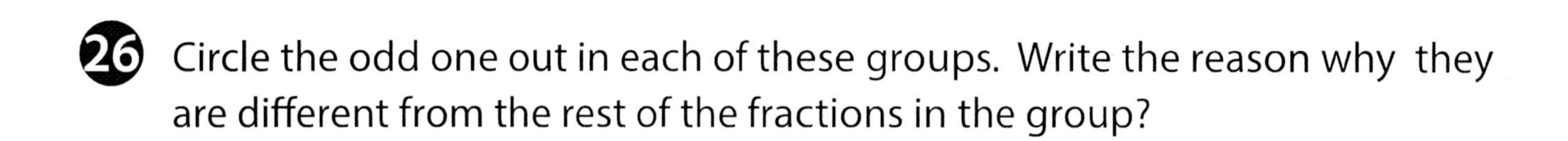

26 Circle the odd one out in each of these groups. Write the reason why they are different from the rest of the fractions in the group?

a $\frac{3}{6}$ $\frac{4}{8}$ $\frac{2}{5}$ $\frac{5}{10}$ ....

....

b $\frac{2}{8}$ $\frac{1}{4}$ $\frac{3}{9}$ $\frac{3}{12}$ ....

....

27 Siddiq ate $\frac{2}{5}$ of his sweets. There were **15** sweets left.
How many sweets did he have before he ate some?

28 Colour the fraction shown for these shapes.

a $\frac{2}{5}$

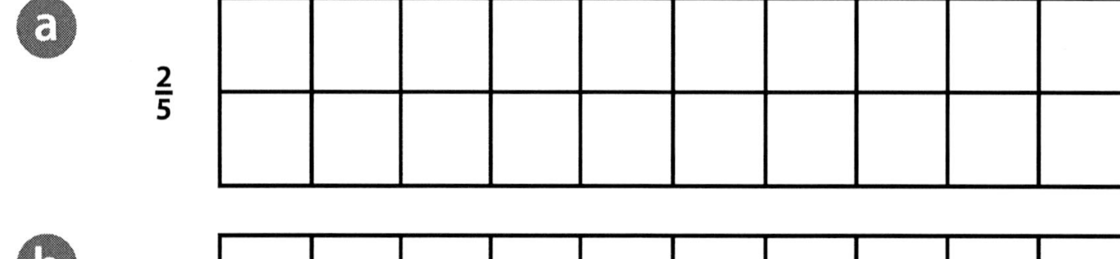

b $\frac{3}{4}$

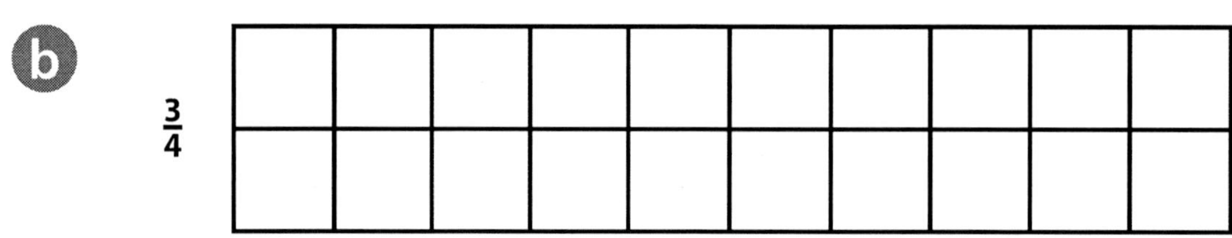

29 Sarah had **24** Top Trump cards. She gave $\frac{2}{3}$ of them to Humma.
How many cards did she have left?

Name ....

30 Choose numbers from the box on the right to make these number sentences correct.

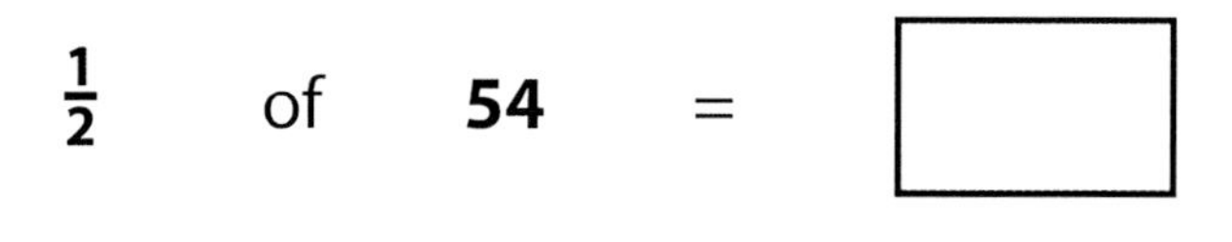

$\frac{1}{2}$ of **54** = ☐

$\frac{1}{4}$ of **36** = ☐

$\frac{1}{3}$ of **48** = ☐

$\frac{1}{8}$ of **56** = ☐

$\frac{1}{10}$ of **220** = ☐

7
22
9
27
16

---

31 **a** Put a circle around $\frac{1}{5}$ of these stars.

b Colour in $\frac{3}{4}$ of the stars that are not circled.

c How many stars are left?

☐

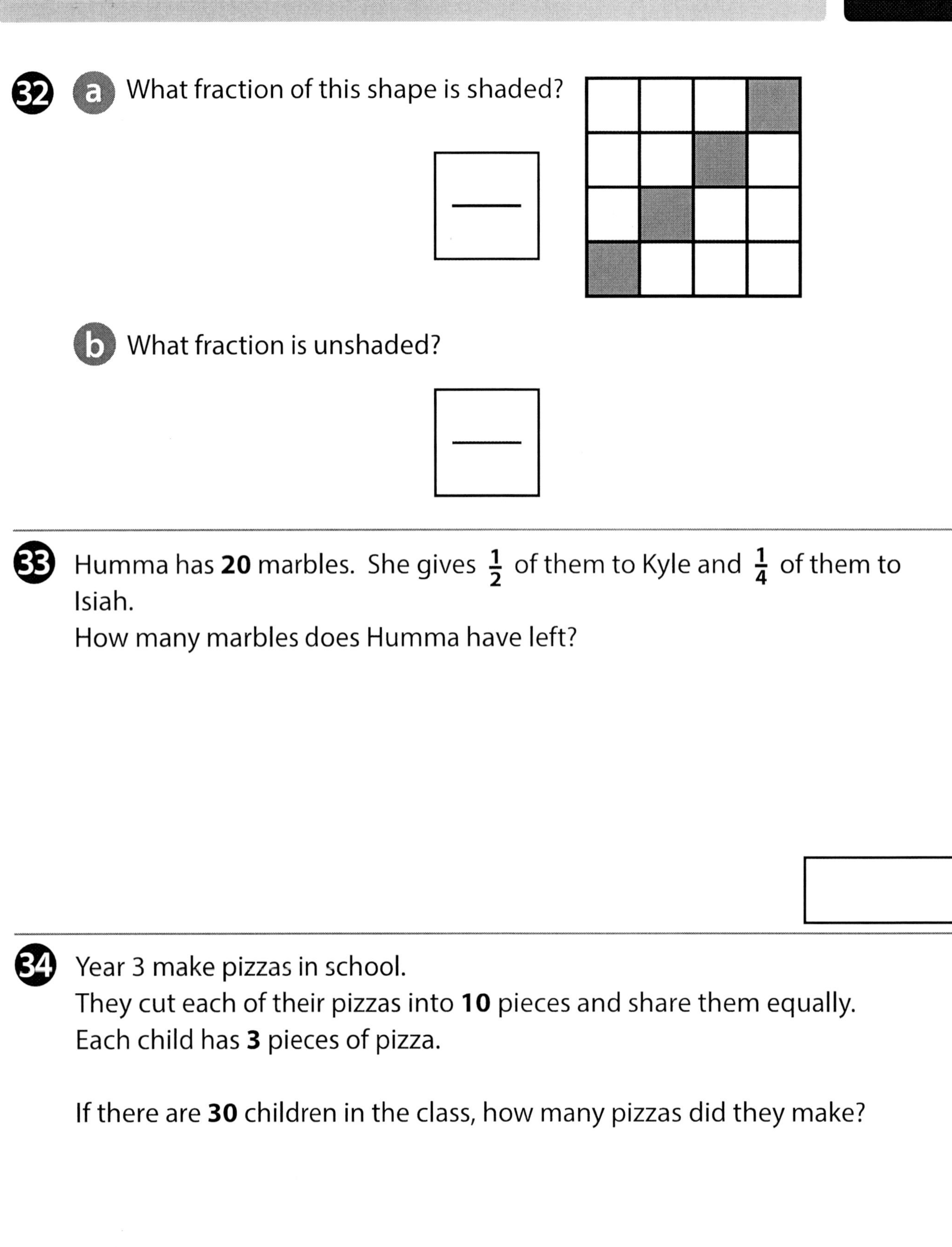

32 **a** What fraction of this shape is shaded?

b What fraction is unshaded?

33 Humma has **20** marbles. She gives $\frac{1}{2}$ of them to Kyle and $\frac{1}{4}$ of them to Isiah.
How many marbles does Humma have left?

34 Year 3 make pizzas in school.
They cut each of their pizzas into **10** pieces and share them equally.
Each child has **3** pieces of pizza.

If there are **30** children in the class, how many pizzas did they make?

Name ....

There are **300** chairs in the hall. Mr Mop, the caretaker, wants to mop the floor.

Year 3 stack $\frac{1}{10}$ of the chairs.

Year 4 stack $\frac{2}{10}$ of the chairs.

Year 5 stack $\frac{3}{10}$ of the chairs.

Year 6 stack $\frac{4}{10}$ of the chairs.

How many chairs are left for the caretaker to stack before he can mop?

Name ....

It may be appropriate for children to use exercise books or paper to record their answers, working out or explanations.

MEASUREMENT

Solve problems involving comparing lengths

1 Kyle had a stick of seaside rock measuring **23 cm** in length. Isiah's stick of rock was **32 cm**. Whose piece of rock was longer?

2 Siddiq's stick measures **45.5 cm**. Kyle's stick measures **44.5 cm**. Whose stick is longer?

3 Sarah's hand span measures **8 cm 5 mm**. Jessica's hand span measures **86 millimetres**. Who has the wider hand span?

4 The height of the dining room door is **2 m 25 cm**. The height of the kitchen door is **2 m 30 cm**. Which door is taller?

5 Kyle is **124 cm** tall. Siddiq is **1 m 16 cm** tall. Who is taller?

6 Humma's sunflower grew **1 m 20 cm** in the month of May. In June, the sunflower grew **115 cm**. In July, the sunflower grew **90 cm**. In which month did Humma's sunflower grow the most and in which month did it grow the least?

most = 　　**least =**

Name ....

Solve problems involving comparing mass (weight)

1. Sarah's chocolate bar weighs **80 grams**. Kyle's chocolate bar weighs **75 grams**. Whose chocolate bar is heavier?

2. Jessica is making a cake. She puts **180 g** of flour into a red bowl and **120 g** of flour into a blue bowl. Which bowl is heavier?

3. A glass-top table in the furniture store weighs **86.5 kg**. A wooden table weighs **85.6 kg**. Which table is heavier?

4. Siddiq weighs **46.4 kg**. His dad weighs **64.4 kg**. Who is heavier?

5. The tiger weighs **202 kilograms**. A tigress weighs **200,000 grams**. Which is heavier?

6. Mrs Jackson's handbag weighs **1.2 kg**. Miss Denby's bag weighs **1250 grams**. Mrs Whitford's bag weighs **1 kg 300 g**. Whose bag is the heaviest and whose bag is the lightest?

heaviest =   **lightest =**

Name ....

Solve problems involving comparing capacity

1. Jessica has a glass containing **78 ml** of orange juice. Humma has a glass containing **87 ml** of orange juice. Whose glass contains more orange juice?

2. Siddiq buys a **$2\frac{1}{2}$ litre** bottle of cola. Humma buys a bottle containing **$2\frac{3}{4}$ litres** of cola. Who has more cola?

3. The red jug holds **500 millilitres** of water. The green jug holds **$\frac{6}{10}$ litres** of water. Which jug has the bigger capacity?

4. At the aquarium, there are **425 litres** of water in Fish Tank **A**. Fish Tank **B** has **452 litres** of water. Which tank has more water?

5. Isiah has a jug filled with **1200 ml** of orange juice. Siddiq has **1 litre 25 ml** of orange juice in his jug. Who has more orange juice?

6. Kyle's paddling pool is filled with **93,000 millilitres** of water. Siddiq's paddling pool is filled with **94 litres** of water. Sarah's paddling pool is filled with **92,000 millilitres** of water. Whose pool has the most and whose pool has the least water?

most =   **least =**

Name ....

Solve problems involving comparing length, mass and capacity

1 In Mr Thackeray's garden, the green barrel can hold **67 litres** of rainwater. The brown barrel can hold **76 litres** of rainwater. Which barrel has the bigger capacity of water?

---

2 Sarah weighs **42.5 kg**. Her mum weighs **45.2 kg**. Who is heavier?

---

3 A garden has **two** fences. Fence **A** is **38 m** and Fence **B** is **3900 cm**. Which fence is longer?

---

4 The distance from Netherton to Brigley is **48 km**. The distance from Netherton to Hoxley is **49,000 m**. Which town is further from Netherton?

---

5 Kyle walked to school every morning. If he walked past the shops, the distance was **221 m**. If he walked past the park, the distance was **212 m**. Which was the shorter way to school? How much shorter?

Which way? **shops / park**

How much shorter? ______ **m**

---

6 Humma is **110 cm** tall, Isiah is **1 m 20 cm** tall and Jessica is **1356 mm** tall. Who is the tallest and who is the shortest?

tallest = ______ **shortest =** ______

Name ....

Solve problems involving adding and subtracting lengths

1. Sarah threw her ball **5 metres**. Then she threw it another **4 metres**. How far did Sarah throw the ball altogether? Convert your answer to **centimetres**.

   ☐ **cm**

2. Jessica is **125 cm** tall. Over the year she grows **16 cm**. How tall is Jessica now?

   ☐ **cm**

3. The bathroom window is **66 cm** long. The bedroom window is **93 cm** long. How much longer is the bedroom window than the bathroom window?

   ☐ **cm**

4. Siddiq is **1 m 42 cm** tall. Jessica is **23 cm** shorter. How tall is Jessica?

   ☐

5. On Thursday, the builder built a wall **53 m** long. On Friday, he built another **17 m** of wall. On Saturday, a truck accidentally knocked down **23 metres** of the wall. How many **metres** of wall were left?

   ☐ **m**

6. A paper chain measures **365 cm** and another paper chain measures **257 cm**. The two paper chains are joined together to make a new one. How much longer than **6 metres** is the new paper chain?

   ☐ **cm**

Name ....

Solve problems involving adding and subtracting mass (weight)

1 A comic weighs **50 grams** and Kyle buys **2** comics. What is the weight of both comics?

☐ g

2 Jessica has **58 grams** of sweets in a paper bag. She eats **23 grams** of the sweets. What weight of sweets does she have left?

☐ g

3 Sarah's empty pencil case weighs **35 grams**. Sarah puts in a pencil sharpener which weighs **25 grams** and a pencil which weighs **15 grams**. What is the weight of the pencil case now?

☐ g

4 To make **2** cakes, Humma needs **110 g** of flour, **55 g** of butter and **50 g** of sugar. What is the weight of all **3** ingredients added together?

☐ g

5 Humma and her mum and dad are going on holiday on a plane. Humma's case weighs **23 kg**. Her mum's case weighs **24 kg** and her dad's case weighs **28 kg**. The weight limit for each case is **20 kg**. How much over the total weight limit are the **3** cases altogether?

☐ kg

6 To bake **one** loaf of bread, Kyle needs **350 grams** of flour. He buys **1 kilogram** of flour. How many loaves of bread can he bake? How much flour will he have left?

loaves ☐ **flour left** ☐ g

Name ....

Solve problems involving adding and subtracting capacity

1 There are **100 millilitres** of milk left in the bottle. Isiah pours **70 millilitres** onto his cereal. How much milk is left?

[ ] **ml**

2 Mrs Johnson is making a cup of tea. She puts in the tea bag then pours **200 ml** of hot water into her cup. She then adds **30 ml** of milk. How many **millilitres** are in the cup altogether?

[ ] **ml**

3 Jessica pours **25 ml** of cordial into a glass. She then adds **75 ml** of water. How much is in the glass altogether?

[ ] **ml**

4 Humma bought **$6\frac{1}{2}$ litres** of fizzy drink for her party. At the end of the party, she had **2 litres** left. How many **litres** of fizzy drink were drunk at her party?

[ ] **litres**

5 Mrs Iqbal has **55 litres** of petrol in her car. She uses **20 litres** and then puts in another **15 litres**. How much petrol is in the car now?

[ ] **litres**

6 The corner shop sold **7 litres** of lemonade in the morning and **10 litres** in the afternoon. There were **6 litres** left. How many **litres** did the shop have to begin with?

[ ] **litres**

Name ....

Solve problems involving adding and subtracting length, mass and capacity

1. A sunflower was **75 centimetres** tall. It grew **9 centimetres** in a month. How tall is the sunflower now?

   ☐ **cm**

2. Sarah has a **500 ml** bottle of juice. She drinks **200 ml**. How much juice does she have left?

   ☐ **ml**

3. Mrs Davies is **1 m 53 cm** tall. She puts on a hat which makes her **9 cm** taller. How tall is she with the hat on?

   ☐ **cm**

4. Siddiq travels **62 km** by coach, **35 km** by taxi and then he walks for **5 km**. How far does he travel in total?

   ☐ **km**

5. Mr Wilson's class are building a fence for the class rabbit. The fence needs to be **385 cm** long. On the first day, they build **140 cm**. On the second day, they build another **125 cm**. How many more **centimetres** do they need to build?

   ☐ **cm**

6. Sarah, Siddiq and Kyle were doing a sponsored walk. Sarah walked **3500 metres**. Siddiq walked **$3\frac{1}{2}$ kilometres**. Kyle walked **2500 metres**. How far did they walk altogether? Give your answer in **kilometres**.

   ☐ **km**

Name ....

Solve problems involving adding and subtracting length, mass and capacity

1 Sarah throws a ball for her dog. She throws it **27 metres**. Humma throws the ball **23 metres**. How far did they throw the ball together?

[ ] m

2 Kyle is **126 cm** tall. Siddiq is **15 cm** smaller than Kyle. How tall is Siddiq?

[ ] cm

3 Isiah has a **1 litre** bottle of water. He drinks **375 ml**. How much water does he have left?

[ ] ml

4 Sarah's plant grew to a height of **500 millimetres**. Kyle's plant grew to a height of **53 centimetres** and Isiah's grew to **67 centimetres**. What is the difference in height, in **centimetres**, between the shortest and tallest plant?

[ ] cm

5 Mr Waddington's class are building a **375 cm** tower from cereal boxes. After the first day, it was **128 cm** tall. On the second day, they built another **112 cm**. How many more **centimetres** of the tower do they need to build?

[ ] cm

6 Humma weighed out **1.2 kg** of sugar onto the scales. She added **550 g** of flour. How many **grams** of sugar and flour were on the scales altogether?

[ ] g

Name ....

Add amounts of money and work out change

1 Jessica has **£1.40**. Her mum gives her **50p**. How much money does she have now?

£ ☐

2 Kyle has **four £1** coins and **seven 5p** coins. How much does he have altogether?

£ ☐

3 Siddiq buys **4** jam doughnuts for **60p** each. How much change does he get from **£3**?

☐ p

4 Isiah buys a sandwich for **£1.20** and a biscuit for **45p**. How much change does he get from **£2**?

☐ p

School Fair Prices	
Key ring – **£1.60**	Ball – **£1.20**
Game – **£2**	Pen set – **£1.70**

5 Sarah has **£5**. She buys a ball and a game. How much change does she have?

£ ☐

6 Humma spends **£4.50** on **3** items. What are they?

☐ ☐ ☐

Name ....

Subtract amounts of money and work out change

1 Isiah has **90p**. He spends **30p** on a bag of crisps. How much money does he have left?

[ ] p

2 Jessica has **£1**. She spends **65p** on an ice cream at Mr Frosty's ice cream van. How much change will she receive?

[ ] p

3 Humma has **£20**. She buys a new jumper for **£9**. She then buys a headband for **£3**. How much money does she have left?

£ [ ]

4 Mr Jones bought **2** plants for his garden. Each plant cost **£1.50**. How much change did he get from **£5**?

£ [ ]

5 In a restaurant, Toffee Dessert costs **£1.75** and Fruit Dessert costs **£2.15**. Humma has a Toffee Dessert and Sarah has a Fruit Dessert. How much change will they have from **£5**?

£ [ ]

6 Jessica has **£5**. A basket of strawberries costs **£1.30**. Does Jessica have enough to buy **4** baskets?

Yes / No

Explain your answer.

....

....

Name ....

Add and subtract money to give amounts of change

1. Sarah has **£2**. She buys a bracelet for **£1**. How much money does she have left?

£ ☐

2. Humma has **£1.40**. Her grandma gives her **30p**. How much money does she have altogether?

£ ☐

3. Kyle has **£1.20**. He wins another **£1.30** playing a fairground game. How much does he have now?

£ ☐

4. Kebabs cost **£1.45** each. Isiah buys **one** and pays with a **£2** coin. How much change does he get?

☐ p

5. What is the total cost of a cupcake costing **35p**, an apple costing **15p** and a drink costing **40p**?

☐ p

6. Humma has saved **three 50p** coins and **four 20p** coins. How much has she saved in total?

£ ☐

7. Siddiq has **four £1** coins, **two 20p** coins and **one 10p** coin. He then spends **50p** on a drink. How much money does he have left?

£ ☐

Name ....

Add and subtract money to give amounts of change

1 Sarah has **105p** and Kyle has **150p**. How much money, in **pence**, have they got altogether?

[ ] p

2 Humma's mum gives her **three £1** coins, **two 5p** coins and **six 10p** coins. How much does Humma's mum give her altogether?

£ [ ]

3 Jessica has **two 50p** coins and **five 20p** coins. Her favourite magazine costs **£1.70**. How much change does she get?

[ ] p

4 Look at the cost of rides at the Funky Fairground:

Big Dipper – **90p**
Space Stroller – **£1.20**
Big Wheel – **£1.30**
Dodgems – **80p**
Twister – **70p**
Ghost Train – **50p**

How much would it cost to go on the Twister and the Space Stroller?

£ [ ]

5 Sarah has **£2**. She goes on **more than one** ride and does not get any change. Which rides could Sarah have gone on?

[ ] **and** [ ]

Name ....

Record and compare time in terms of seconds, minutes and hours and o'clock

1. It took Kyle just **two minutes** to score his first goal of the match. How many **seconds** was this?

   [ ] **seconds**

2. Isiah ran one lap of the track in **3 minutes**. Siddiq ran one lap in **190 seconds**. Who was faster? By how many **seconds**?

   [ ] **by** [ ] **seconds**

3. Kyle went swimming for **one hour**. Sarah went swimming for **60 minutes.** Did they go swimming for the same amount of time?

   Yes / No

   Explain your answer.

   ....

   ....

4. Isiah and Jessica went to the cinema but watched different films. Isiah's film lasted **one hour, forty five minutes**. Jessica's film started at **6.40 pm** and lasted until **8.15 pm**. Whose film was longer?

   Isiah's / Jessica's

   Explain your answer.

   ....

   ....

5. The aeroplane to Greece took off at **1 o'clock** and landed at **5 o'clock**. How many minutes did the flight take?

   [ ] **minutes**

Name ....

Use vocabulary such as am/pm, morning, afternoon, evening, noon and midnight

1. Humma went fishing at **2 o'clock** in the afternoon. She told her mum that she would be back by **5 am** on the same day. Why was this impossible?

....

....

2. Sarah is going on holiday. She needs to wake up at **4 am**. She thinks she will be getting up in the afternoon. Is she correct?

Yes / No

Explain your answer.

....

....

3. Jessica has a picnic at **12 noon**. Her friend says that it will be dark at that time. Is her friend correct?

Yes / No

Explain your answer.

....

....

4. Siddiq is going to a party at **6 pm**. He tells his mum he will be back at **9 am** that evening. Is he correct?

Yes / No

Explain your answer.

....

....

Name ....

Know the number of seconds in a minute

1. Kyle held his breath for **one minute**. How many **seconds** was this?

[ ] **seconds**

2. Jessica wanted to skip for **one minute** without stopping. If she had already skipped for **45 seconds**, how many more **seconds** did she need to skip for?

[ ] **seconds**

3. Year 3 were catching the bus to the swimming pool in **2 minutes**. How many **seconds** did they have to wait to catch the bus?

[ ] **seconds**

4. **90 seconds** were added on to the end of the match as additional time. How many **minutes** was this?

[ ] **minutes**

5. Humma's mum said she had to go to bed in **5 minutes**. How many **seconds** did she have left to stay up?

[ ] **seconds**

6. Which is longer: **240 seconds** or **3 minutes**? Explain your answer.

[ ] because ....

....

Name ....

Know the number of days in each month

1 How many **days** are there in **January**?

☐ **days**

2 Which has more **days**: **September** or **July**?

☐

3 Put the months below in order starting from the month with the least **days**, to the month with the most.

May **February** **November**

☐ ☐ ☐

4 Isiah was going on a long trip for all of **April** and **May**. How many **days** was he going for altogther?

☐ **days**

5 How many **days** are in **February** and **March** altogether, if it is a leap year?

☐ **days**

6 How many **days** are there altogether in **June**, **July**, **August** and **September**?

☐ **days**

Name ....

Know the number of days in a year and a leap year

1 Sarah went running every day for **one year** (not a leap year).
For how many **days** did she go running?

☐ **days**

---

2 It's Jessica's birthday on the **29th February**. How many **years** will it be until she next celebrates her birthday on the **29th February**?

☐ **years**

---

3 How many **days** are there in a **leap year**?

☐ **days**

---

4 In 2017 (not a leap year), Siddiq lived in England but went to Italy to stay with his cousins for **June** and **July**. For how many **days** was Siddiq in England during 2017?

☐ **days**

---

5 How many **days** are there in **July**, **August** and **October** altogether?

☐ **days**

---

6 Humma lives in England but, in 2017, she went on holiday to Australia for **6 weeks**. How many **days** was Humma in England during 2017?

☐ **days**

Name ....

Calculate the time taken by particular events

1 In a race, Sarah ran **2** laps of the school field. It took her **two minutes**. How many **seconds** did it take her?

[ ] **seconds**

2 Humma and Sarah went swimming at **12 noon**. They stayed in the pool for **two hours**. What time did they finish swimming?

[ ]

3 The time on the clock showed **11:15**. The correct time was **11:21**. How many **minutes** slow was the clock?

[ ] **minutes**

4 Jessica goes running at **3.30 pm**. She runs for **40 minutes**. What time does she finish running?

[ ]

5 School lunch lasts for **1 hour 10 minutes**. It ends at **1.10 pm**. What time does it start? Use morning, afternoon, noon or midnight with your answer.

[ ]

6 The train to the seaside leaves the station at **twenty to eight** in the morning. It arrives at the seaside at **9.30 am**. How long does the journey take?

[ ]

Name ....

Calculate the time taken by particular events

1. A film lasted for an **hour**. For how many **minutes** did the film last?

☐ **minutes**

2. Humma went running at **10.20 am**. She returned at **10.45 am**. How long did Humma run for?

☐ **minutes**

3. Siddiq went to the mosque at **4.30 pm**. He left at **5.20 pm**. How long did he stay at the mosque?

☐ **minutes**

4. Kyle went to bed when his clock read **15 minutes past 8**. He read until **8.40 pm**. For how long did Kyle read?

☐ **minutes**

5. Jessica fell asleep at **9.15 pm**. She woke up at **4.15 am**. How long did Jessica sleep for?

☐ **hours**

6. Kyle played cricket from **1.30 pm** until **2.15 pm**. Siddiq played for **half an hour**. How much longer did Kyle play cricket than Siddiq?

☐ **minutes**

Name ....

Calculate the time taken by particular events

1. Isiah started his maths lesson at **10 am**. It lasted for **50 minutes**. What time did it finish?

2. A bus takes **half an hour** to reach its destination. If it departs at **half past 8** in the morning, what time does it arrive?

3. Kyle started watching a film at **7.30 pm**. The film lasted for an **hour and a quarter**. What time did the film finish?

4. Humma takes **25 minutes** to eat her packed lunch. If she started eating at **12.15 pm**, what time would she finish?

5. Jessica's family set off for Funland Theme Park at **10.50 am**. The journey took **1 hour and 25 minutes**. What time did they arrive at the theme park?

6. Mr Young set off in his car at **3.10 pm**. He drove for **25 minutes** to pick up his daughter. He then drove for **35 minutes** to the park. What time did he arrive at the park?

Name ....

Compare the duration of events

1. Humma's favourite TV programme lasted for **35 minutes**. Isiah's lasted **half an hour**. Whose was shorter? By how much?

   [ ] **by** [ ] **minutes**

2. Sarah and Humma were running in a race. Sarah finished the race in **12 minutes**. Humma took a **quarter of an hour** to finish. Who finished the race **first**?

   [ ]

3. Jessica and Isiah started their homework at **4.15 pm**. Jessica finished in **45 minutes**. Isiah finished at **4.55 pm**. Who finished their homework **first**?

   [ ]

4. The train to the city takes **1 hour and 12 minutes**. The train to the countryside takes **88 minutes**. Which journey is longer?

   [ ]

5. In Class 3, on a Tuesday, they have maths from **9.10 am** until break at **10.00 am**. After break, they have English from **11.00 am** until **11.55 am**. Which lesson is longer: English or maths?

   [ ]

6. Isiah completed the running section of the triathalon from **3 pm** until **4.30 pm**. He completed the swimming section from **4.30 pm** until **5.15 pm**. Which took him longer: running or swimming?

   [ ]

Name ....

It may be appropriate for children to use exercise books or paper to record their answers, working out or explanations.

MASTERING

Measurement

1 I have a **5 m** roll of border.
The display boards in Year 3 measure **120 cm** wide and **70 cm** high.

a How much border do I need to go around **one** display board?

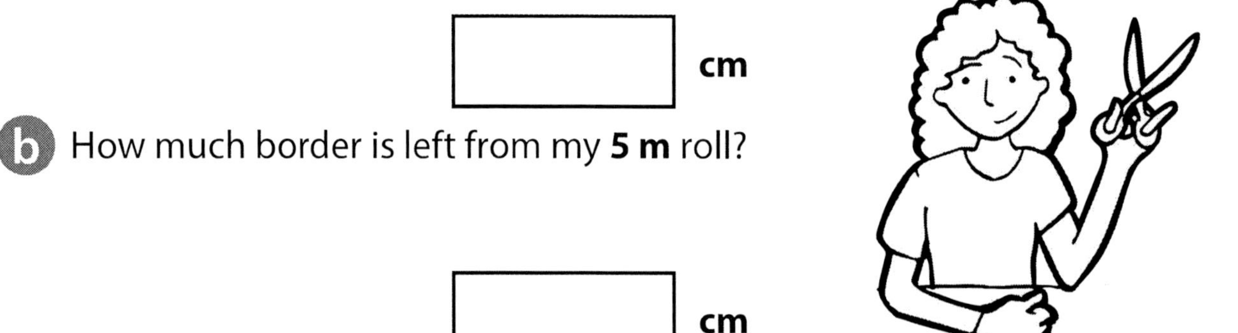

cm

b How much border is left from my **5 m** roll?

cm

c I have **3** display boards. How many rolls of border do I need?

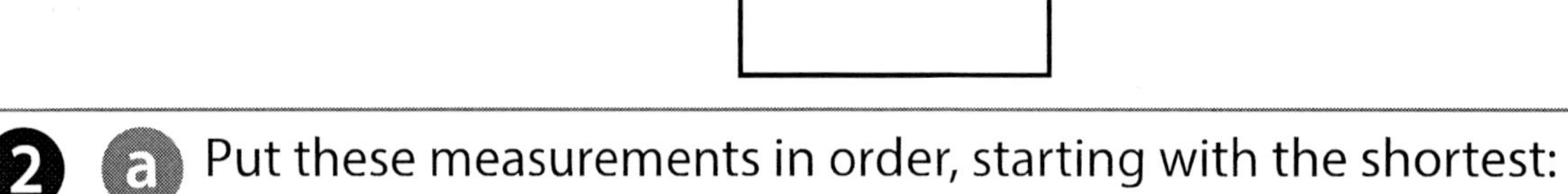

---

2 **a** Put these measurements in order, starting with the shortest:

27 cm $\frac{1}{4}$ **m** **260 mm**

b What is the difference between the longest and shortest lengths?
Give your answer first in **centimetres** and then in **millimetres**.

cm mm

Name ....

3 Sarah's pencil is twice as long as Humma's pencil.
Jessica's pencil is $2\frac{1}{2}$ **cm** longer than Sarah's pencil.
Jessica's pencil is $12\frac{1}{2}$ **cm** long.
How long are Sarah and Humma's pencils?

Sarah's pencil ☐ **cm** **Humma's pencil** ☐ **cm**

4 Use the symbols **<, >, =** to compare these measurements:

$\frac{1}{2}$ **litre** ☐ **450 ml**

50 cm ☐ $\frac{1}{2}$ **m**

700 g ☐ $\frac{1}{2}$ **kg**

5 **a** Find the weight of a banana.

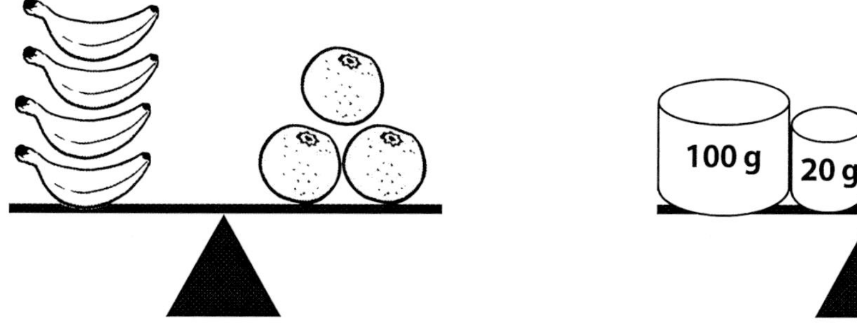

1 banana = ☐ **g**

b Now find the weight of the pineapple.

☐ **g**

Name ....

6 Kyle starts reading a book at **2.30 pm**. He reads for $\frac{1}{2}$ an hour.
He says, "I have read my book for **50 minutes**!"
Is Kyle correct?

Yes / No

Explain your answer.

....

....

---

7 Siddiq has a **2 litre** bottle of lemonade. He drinks $\frac{1}{2}$ **litre**.

a What fraction of the lemonade is left?

$\frac{\square}{\square}$

b What fraction of the lemonade is left after Siddiq has drunk $\frac{3}{4}$ **litre**?

$\frac{\square}{\square}$

c How many **millilitres** are left in the bottle after Siddiq has drunk $\frac{3}{4}$ of the lemonade?

☐ **ml**

---

8 Sarah wants to make orange squash for herself and **six** friends.
Each child needs **150 ml** of orange squash.
Sarah has a jug that can hold **1 litre**.
Is this jug large enough to hold all the orange squash that Sarah needs?

Yes / No

Explain your answer.

....

....

Name ....

9 Isiah and Kyle have saved some money.
Altogether they have saved **£65**. Kyle has saved **£15** more than Isiah.
How much have they each saved?

Isiah £ ☐ **Kyle £** ☐

---

10 Jessica draws a square and Siddiq draws a rectangle.
Both shapes have sides that measure a whole number of centimetres.
The perimeter of the square is the same as the perimeter of the rectangle.

a If the perimeter of the square is **16 cm**, how long is each side?

☐ **cm**

b What could the measurements of the rectangle be?

☐ **cm by** ☐ **cm** **or** ☐ **cm by** ☐ **cm**

or ☐ **cm by** ☐ **cm**

c The length of the rectangle is **three** times the width of the rectangle.
What are the actual measurements of the rectangle that Siddiq drew?

☐ **cm by** ☐ **cm**

Name ....

11 Isiah bought a magazine and a book and paid with a **£10** note.
He received **2** silver coins as change.

a Which coins could he have received?

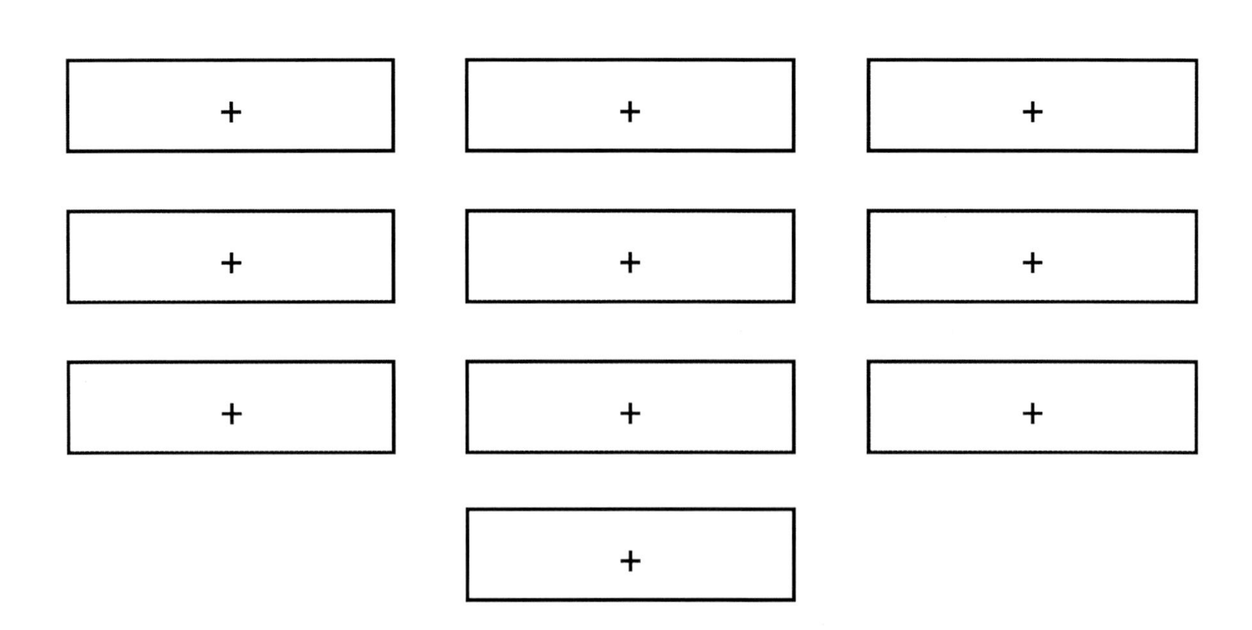

b The **two** coins in Isiah's change were not the same as each other.
The larger coin was **ten** times the value of the smaller coin.
How much change did Isiah receive?

c The book cost exactly **twice** as much as the magazine.
How much did Isiah pay for the book and the magazine?

book £

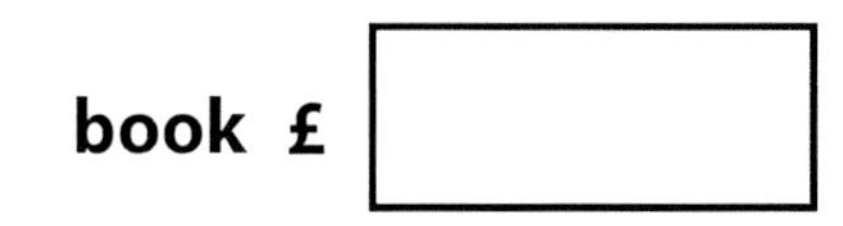

magazine £

Name ....

12 Siddiq buys **2** pencils and a pen and receives **20p** change from a **£5** note.
The pen costs **twice** as much as each pencil.
How much does a pencil and a pen cost?

pencil £ [ ]

pen £ [ ]

---

13 These clocks have only an hour hand.
Suggest the time that they could be showing. Draw in the minute hands.

a
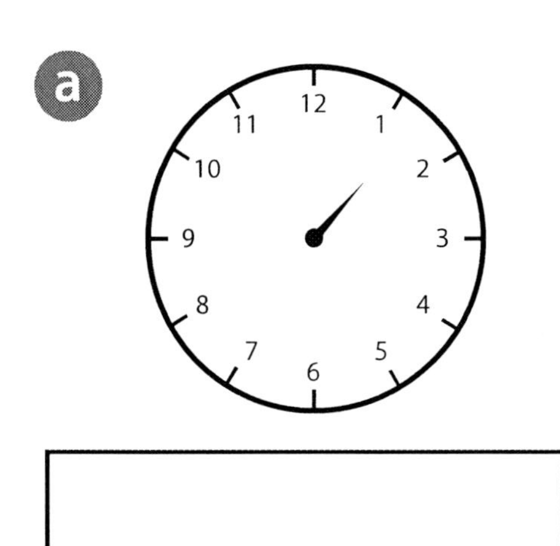

[ ]

b
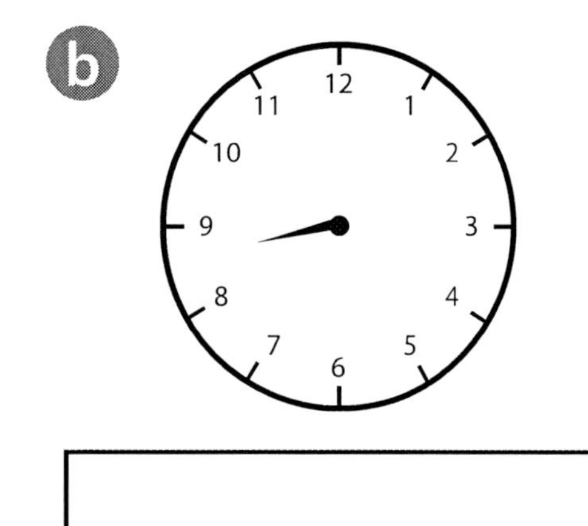

[ ]

c
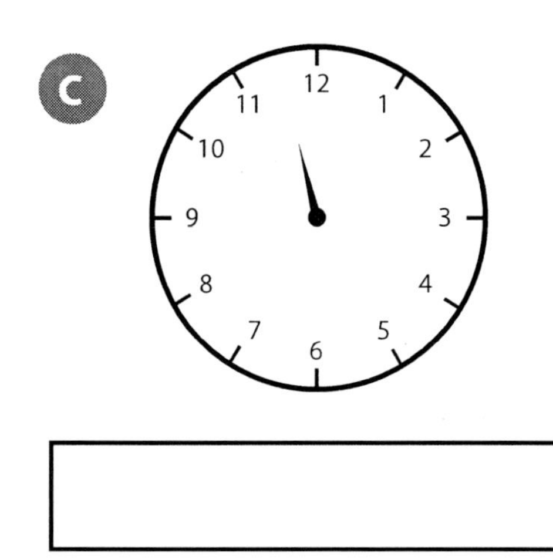

[ ]

d
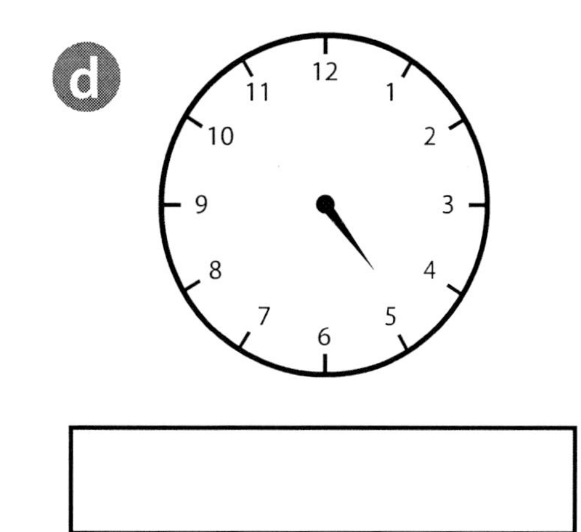

[ ]

Name ....

14 Trains from Ainford to Farthington are shown in this timetable.

Ainford	11:00	12:00	1:30	2:47
Bamwood	11:13			
Catchpole	11:25			
Dundermouth	11:37			
Elderton	11:50			
Farthington	12:00			

a How many trains are shown here?

The time between stops is the same for each train.

b Complete the timetable.

You are in Catchpole and you want to get to Farthington by **2 pm**.

c What is the time of the train that you need to catch at Catchpole?

d How long is your journey on the train?

minutes

Name ....

15 **One** length of a swimming pool is **22 m**. Sarah swims **8** lengths.

a How far does Sarah swim altogether?

[ ] **m**

b Sarah decides to swim the same distance every day after school.
How far does Sarah swim in **one** week?

[ ] **m**

c How much further does Sarah need to swim in order to swim **one kilometre**?

[ ] **m**

16 Humma has these coins in her pocket.
Show which amounts she could make using only **two** coins.
The first **one** has been done for you.

50p + 20p = 70p		

Name ....

17 Siddiq drives to work on Monday. He leaves at **7.35 am** and arrives at **8.15 am**.

a How long does his journey to work take?

☐ **minutes**

On Tuesday, Siddiq leaves at the same time, but drives to the supermarket before driving to work.
He arrives at work at **8:25**.

b How much longer does Siddiq's journey to work take on Tuesday than on Monday?

☐ **minutes**

It took Siddiq **36** minutes to drive to the supermarket.

c How long did it take Siddiq to drive from the supermarket to work?

☐ **minutes**

On Wednesday, Siddiq needs to be at work by **07:55**.

d At what time should Siddiq leave home in order to arrive at work by **7.55 am**?

☐

Name ....

18 Humma has entered a cross-country race.
She needs to run around the school playing field **4** times.
Each lap of the playing field takes **2 minutes 35 seconds**.

a How long should it take Humma to run around the playing field **four** times if she ran at the same speed?

During the race, Humma ran the first lap in **2 minutes 35 seconds**, and then each of the other laps **5 seconds** slower than the previous lap.

b How long did it actually take Humma to run the race?

Here is a list of the times for different children in the race.

c Write the names of the runners showing the order in which they finished the race. Remember to add Humma's name to your list. In which position did Humma finish?

Name	Time
Sarah	11 minutes 5 seconds
Kyle	10 minutes 45 seconds
Siddiq	11 minutes 10 seconds
Isiah	10 minutes 55 seconds

Position	Name
1	
2	
3	
4	
5	

Name ....

19 Isiah is making a cake.
He adds each ingredient to the scales in the order shown below.
He makes a note of the reading each time he adds a new ingredient.

a Fill in the table to show how much each ingredient weighed.

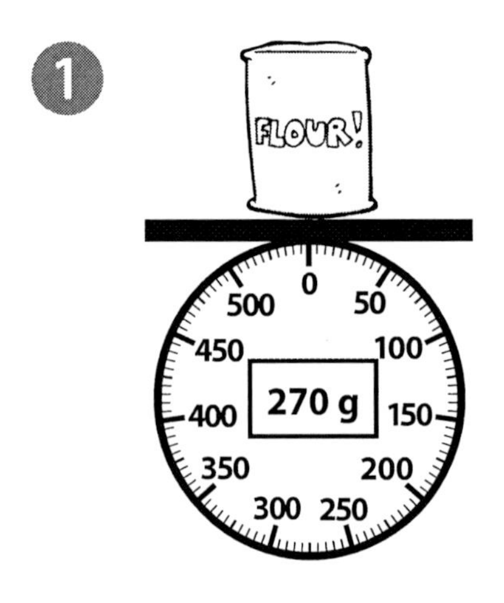

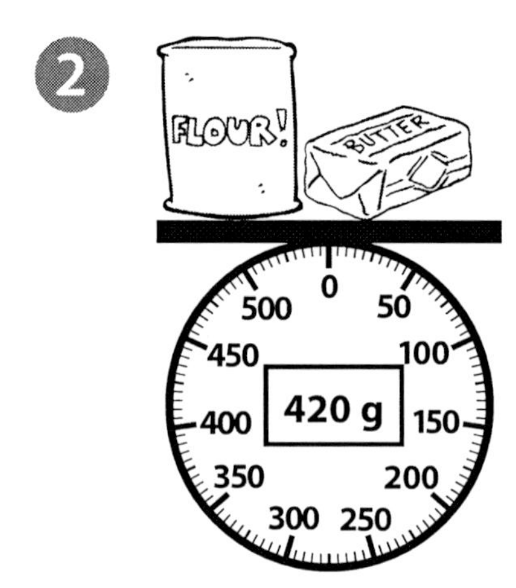

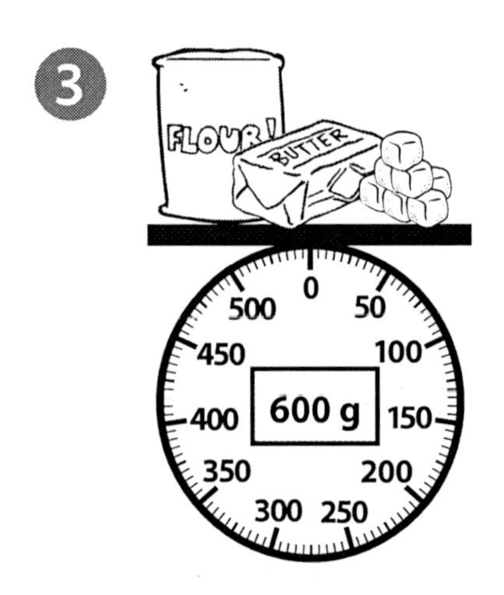

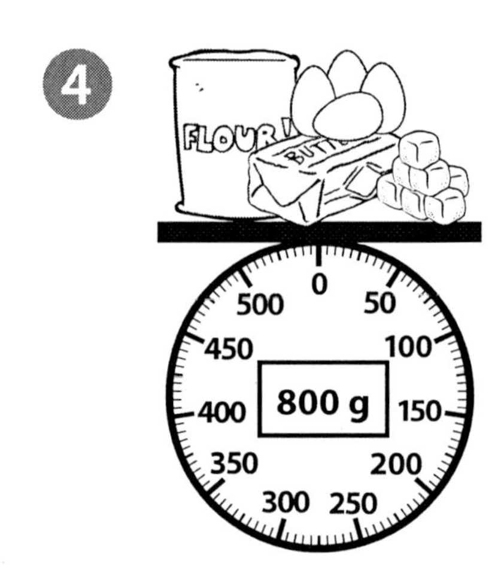

flour	butter	sugar	4 eggs
270 g			

b Isiah also needed to make a cake for Sarah.
Fill in this table to show how much of each ingredient Isiah needs to make both cakes.

flour	butter	sugar	eggs

c Isiah does not have any eggs to make the cakes.
He goes to the shops to buy some.
He can buy either **two** boxes of $\frac{1}{2}$ dozen eggs for **£1.30** each or one box of a dozen eggs for **£2.40**.

How much does Isiah save if he buys **one** box of a **dozen** eggs?

[ ] **p**

d Isiah has a **£5** note.
He needs to buy a newspaper for his dad which costs **70p** and a magazine for his mum that costs **£1.60**.

Does Isiah have enough money for the eggs, the newspaper and the magazine? **Yes / No**

Explain your answer.

....

....

Name ....

It may be appropriate for children to use exercise books or paper to record their answers, working out or explanations.

GEOMETRY

Properties of shapes

Describe and classify 2D and 3D shapes

1 Siddiq draws a **quadrilateral**. How many sides does his shape have?

2 Isiah says a **pentagon** has **6** sides. Is he correct? Explain your answer.

Yes / No

....

....

3 How many sides do **two hexagons** have altogether?

4 Kyle's homework is to draw **2** different types of **triangle**. Describe the **triangles** that Kyle could draw using words or drawings or both.

....

....

5 What is the sum of the number of sides on a **hexagon** added to the number of sides on an **octagon**?

6 Sarah says that her cereal box is a **cuboid**. Name **two** other everyday objects that are **cuboids**.

....

Name ....

Describe and classify 2D and 3D shapes

1 To draw a **cuboid**, would you need to use straight or curved lines?

....

2 What is the name of a **3D** shape which best describes a football?

....

3 What answer would you get if you added the number of sides on a **quadrilateral** to the number of sides on an **octagon**?

☐

4 Draw a **2D** shape that has at least **one** line of symmetry.
Use paper or your book.

5 Isiah names **two 2D** shapes that have **four** right angles each. What might they be?

.... **and** ....

6 Can you name **two 3D** shapes that have **eight** vertices?

.... **and** ....

Name ....

Describe and classify 2D and 3D shapes

1 Siddiq says a **heptagon** has **6** sides. Is he correct? **Yes / No**
Explain your answer.

....

....

2 A **triangular prism** has **5** faces. **3** of the faces are the same shape. What shape is this?

....

3 I am a **2D** shape. I have no straight sides. What shape am I?

....

4 If you added together the number of sides on a **triangle**, a **quadrilateral** and a **pentagon**, how many sides would there be altogether?

5 A **sphere** is a solid shape. A ball is a **sphere**. Can you explain why a **sphere** is a good shape for a ball?

....

....

6 A **cube** is a solid shape. A dice is a **cube**. Can you explain why a **cube** is a good shape for a dice?

....

....

Name ....

Recognise angles as a property of shape and connect right angles and amount of turn

1 Isiah writes the name of a shape that always has **four** right angles. What shape might Isiah write?

---

2 Kyle's class are making a shape dictionary. Complete Kyle's description of a right-angled triangle.

"It has **3** sides and **three** angles.

One of the angles is a ....

---

3 The minute hand of a clock was pointing at **12**. Isiah moved it to point at **3**. What angle had the clock turned through?

---

4 Kyle is facing the door. He makes a **half turn** to face the window. How many right angles has he turned through?

---

5 The minute hand of a clock turns from the number **3** to the number **5**. Jessica said that it had turned more than a right angle. Is she correct?

Explain your answer.

Yes / No

....

....

---

6 Sarah did **5 full turns** very quickly in her ballet class. How many right angles did she turn through altogether?

Name ....

Identify horizontal and vertical lines and pairs of perpendicular and parallel lines

1 To draw a **rectangle**, would you use straight or curved lines?

---

2 Siddiq draws **2** lines. The lines are the same distance apart all the way along their length. What kind of lines has he drawn?

---

3 Mr Little draws this line on the blackboard.

Has he drawn a **horizontal** or a **vertical** line? Explain your answer.

....

....

---

4 How many pairs of **parallel** lines does a **rectangle** have?

---

5 Why would it not be possible for a train to travel along tracks that were not **parallel**?

....

....

---

6 Explain the difference between **perpendicular** lines and **parallel** lines.

....

....

Name ....

It may be appropriate for children to use exercise books or paper to record their answers, working out or explanations.

MASTERING

Geometry

1 Using the dots below draw the following shapes. Some may not be possible.
If they are not possible, explain why not on the lines below.

a a triangle with **one** right angle. **possible / not possible**

b a quadrilateral with **3** right angles. **possible / not possible**

c a quadrilateral with **2** right angles. **possible / not possible**

d a triangle with **two** right angles. **possible / not possible**

....

....

....

Name ....

2 Draw lines to match these **2D** shapes with their names.

pentagon rectangle octagon isosceles triangle

hexagon square parallelogram

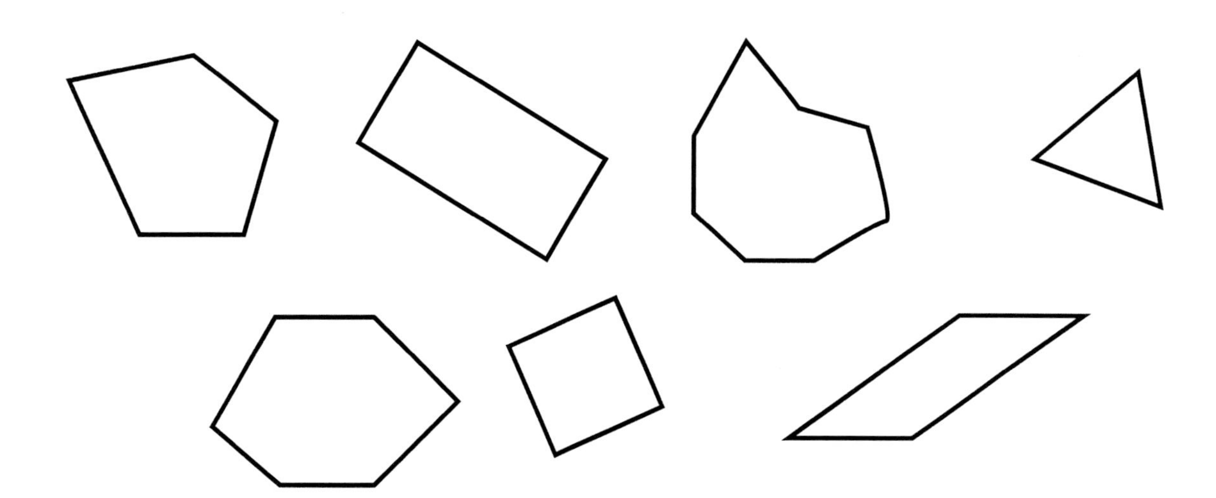

Write the names into the sorting diagram.

	At least one line of symmetry	**No lines of symmetry**
quadrilateral		
not a quadrilateral		

Name ....

3 Kyle draws a square.
Jessica tries to draw the same square, but makes **two** of the opposite parallel sides **3 cm** too short.
Jessica's shape has a perimeter of **58 cm**.

a What is the perimeter of Kyle's square?

cm

b What is the name of Jessica's shape?

c What are the lengths of each of the sides in Kyle's shape?

cm

d What are the lengths of each of the sides in Jessica's shape?

cm cm cm cm

4 Siddiq is facing **east**.
He turns **anti-clockwise** through **2** right angles and then **clockwise** through **3** right angles.

In which direction is Siddiq facing now?

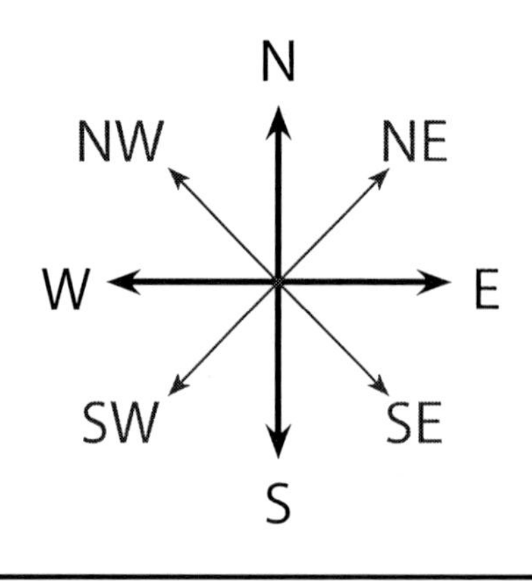

Name ....

5 **a** Sarah is facing west.
She turns **clockwise** through **3** right angles
and then **clockwise** through **2** right angles.

In which direction is Sarah facing now?

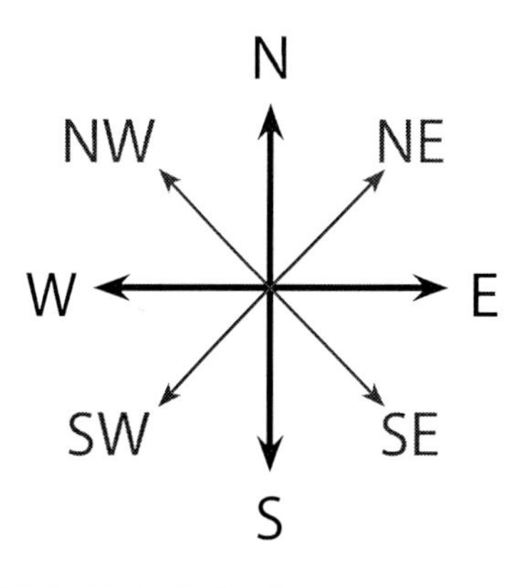

b Jessica is facing **east**.
Through how many right angles **clockwise** does she have to turn in order to be facing **north**?

6 Isiah says that the sides of a hexagon are always the same length.
Is he correct? **Yes / No**

Explain your answer.

....

....

7 Draw all the lines of symmetry on these letters.
Use a ruler.

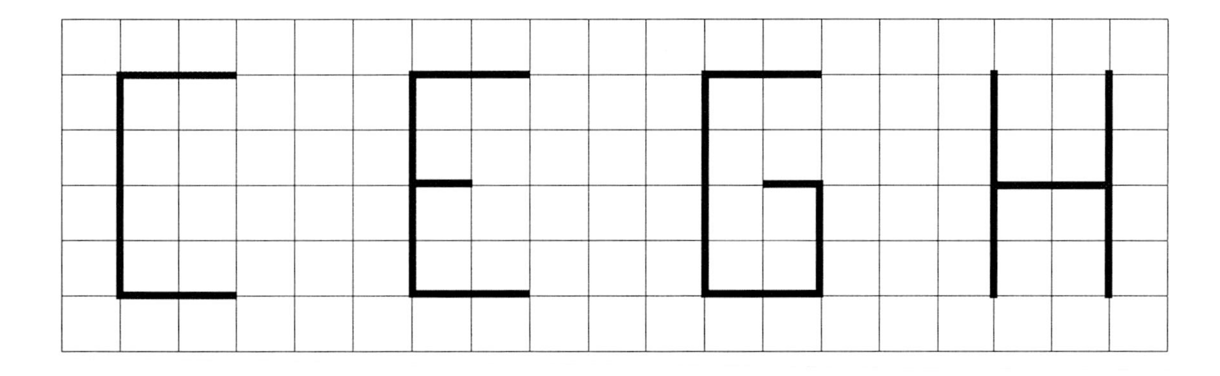

Name ....

8 Look at these shapes.

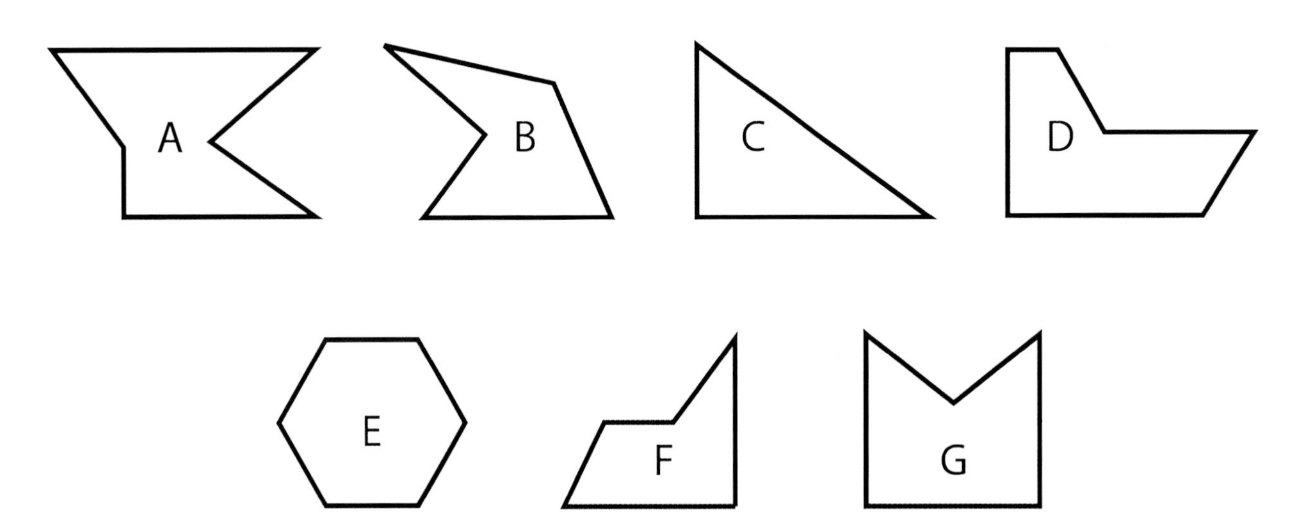

Complete the table by writing the letters of the shapes in the correct places.
You can use a letter more than once.

At least one right angle	At least one pair of parallel sides	Regular shape

Name ....

9 **a** Use a right angle measurer to decide which of these angles are greater than a right angle, and which are smaller than a right angle.

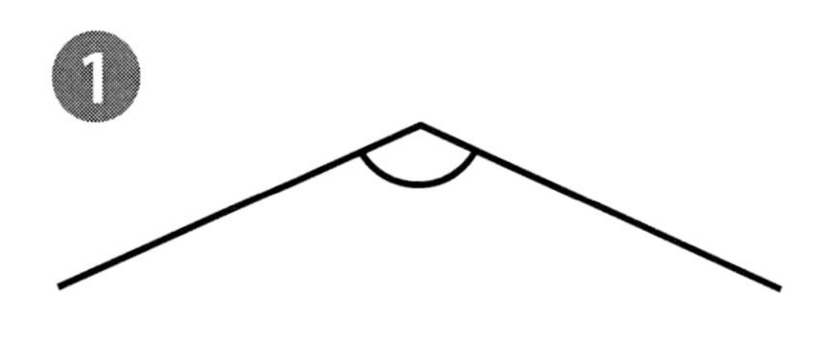

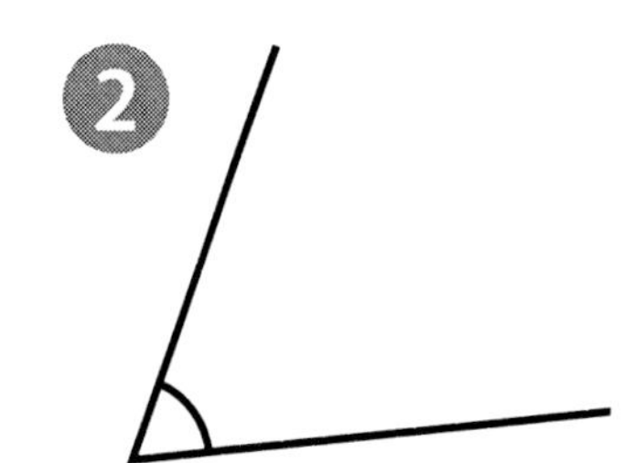

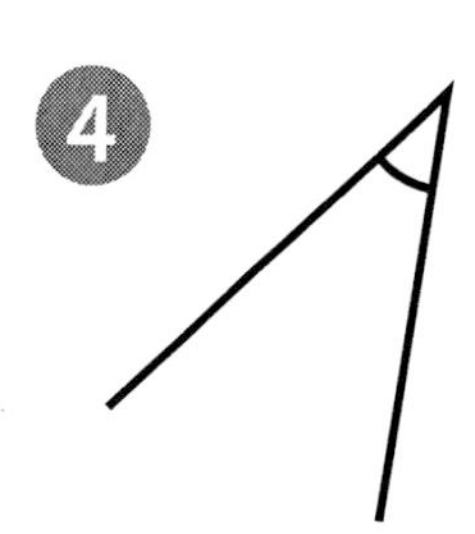

Larger than a right angle	Smaller than a right angle

b Order the angles from smallest to largest. Write the numbers in the boxes below.

smallest **largest**

Name ....

10 **a** Siddiq wants to draw a rectangle.
He knows that:

- the longer sides are **twice** as long as the shorter sides.

- the perimeter of the rectangle is **36 cm**.

Calculate the lengths of the sides and then draw the rectangle using a ruler.
Remember to label the sides of your rectangle with the correct lengths.

b Write down the perimeters of 3 rectangles where the longer sides could be twice as long as the shorter sides.

	Long side	**Short side**	**Perimeter**
1			
2			
3			

Name ....

11 In the word **MATHS** below, the parallel lines and the perpendicular lines have been identified using arrows and the right angle symbol.

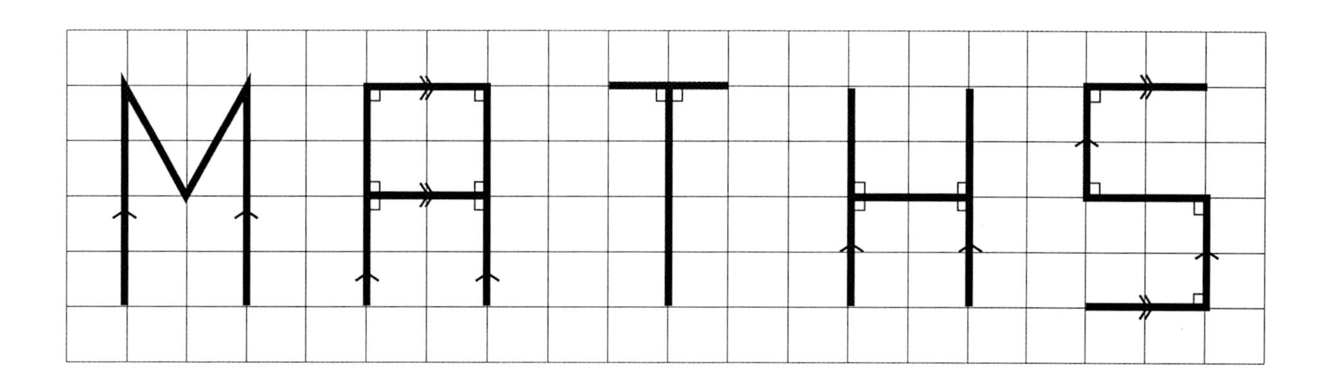

a Write the word **LUNCH** on the grid below and show the parallel and perpendicular lines.

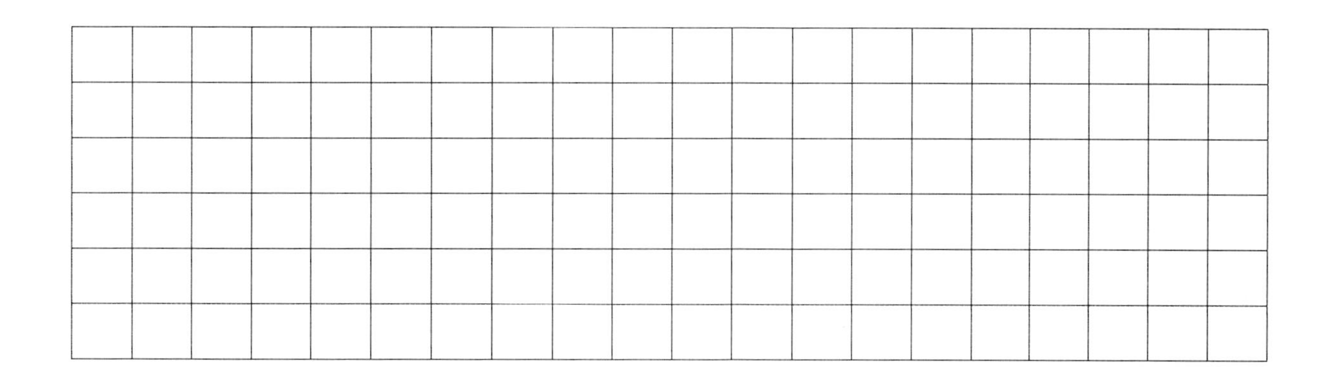

b Now write the word **BREAK** on the grid below and show the parallel and perpendicular lines.

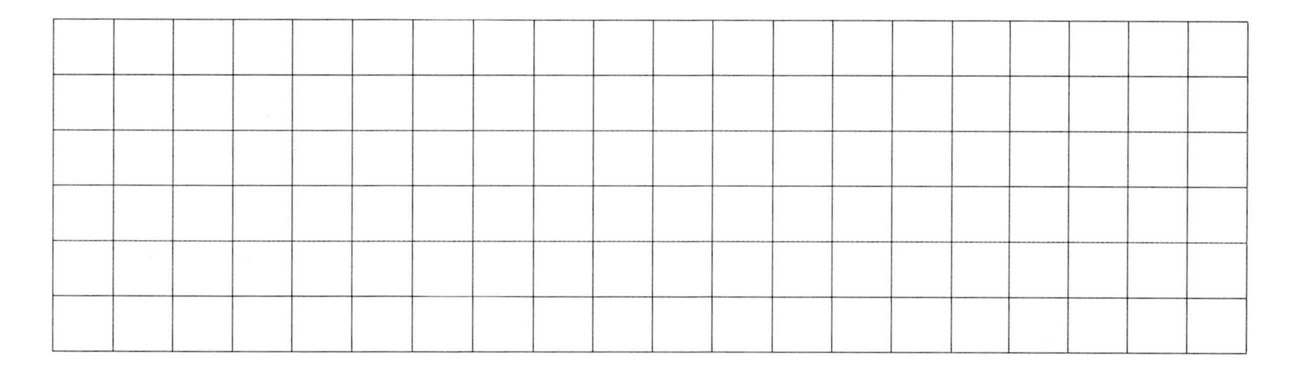

 On the grid below, join dots to make a triangle that does not have a right angle.

 Write the missing numbers in this table.

	square faces	rectangular faces	triangular faces	circular faces
cuboid				
square based pyramid				
cylinder				
triangular prism				
cube				

Name ....

14 Look at the shapes drawn on the grid.

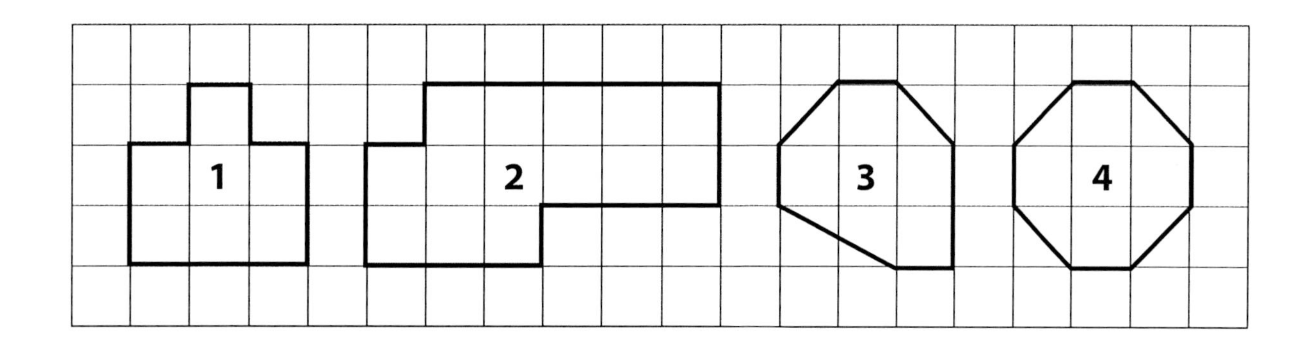

Now complete the table below by ticking the boxes.

	Is an octagon	Has at least one right angle
1		
2		
3		
4		

15 Join the dots to make a regular hexagon.

Name ....

16

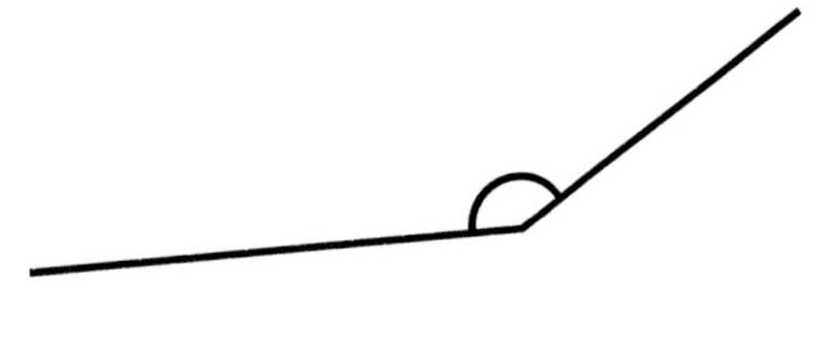

1

3

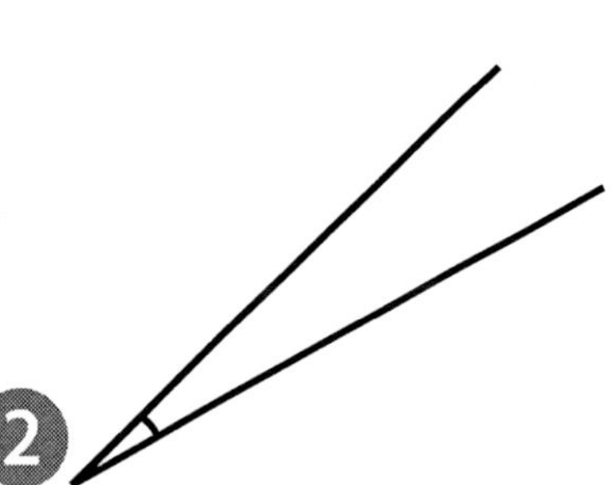

2

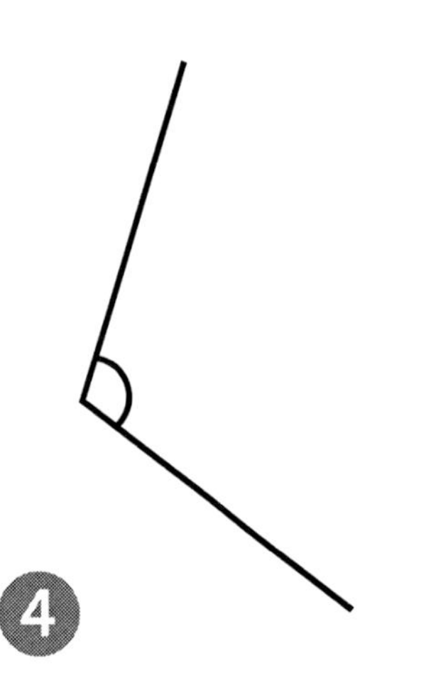

4

5

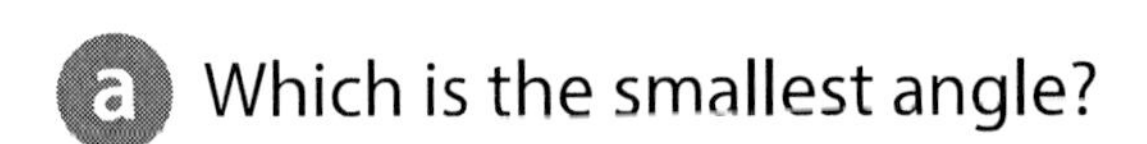

a Which is the smallest angle?

b Which **two** angles are the same size?

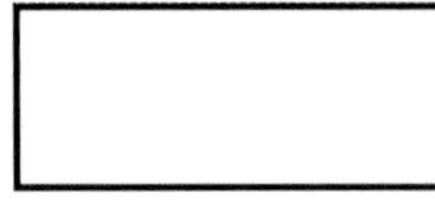

and

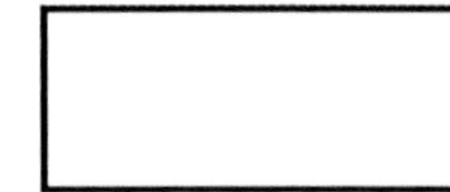

c Which angles are obtuse angles?

and

Name ....

It may be appropriate for children to use exercise books or paper to record their answers, working out or explanations.

STATISTICS

Interpret data and solve problems from a tally chart

Tally chart to show the number of children on the playground at different times

Time	10 am	11 am	12 noon	1 pm	2 pm
Tally	卌 |||	卌		|||	卌 ||||	卌卌 |||

1. When are there the most children on the playground?

2. How many more children were on the playground at **11 am** than at **12 noon**?

3. Why do you think there were the least number of children on the playground at **12 noon**?

....

....

4. How many children were in the playground in total during the afternoon (including those in at **12 noon**)?

5. Sarah said, "There were **twice** as many children on the playground at **2 pm** as there were at **11 am**." Is she correct?

Yes / No

Explain your answer.

....

....

Name ....

Interpret data and solve problems from a tally chart

Tally chart to show the number of animals seen in the park on Tuesday afternoon

Animal	Tally
birds 	~~IIII~~ III
rabbits 	III
dogs	~~IIII~~ I
cats 	
worms	~~IIII~~ ~~IIII~~ III

1. How many birds were there in the park?

2. How many cats were there in the park?

3. How many more birds were there than rabbits?

4. How many more worms were there than dogs?

5. How many creatures were counted in total in the park?

Name ....

Interpret data and solve problems from a tally chart

Tally chart to show the number of animals on the farm

Animal	Tally
horses 	卌 ||
cows 	|||
sheep	卌
pigs	卌 ||||

1 How many cows are on the farm?

2 How many pigs are on the farm?

3 How many more horses are there than cows?

4 Jessica says, "There are more pigs than there are cows and sheep put together." Is this true?

Yes / No

Explain how you know.

....

....

5 How many animals are there in total?

Name ....

Interpret data and solve problems from a bar chart

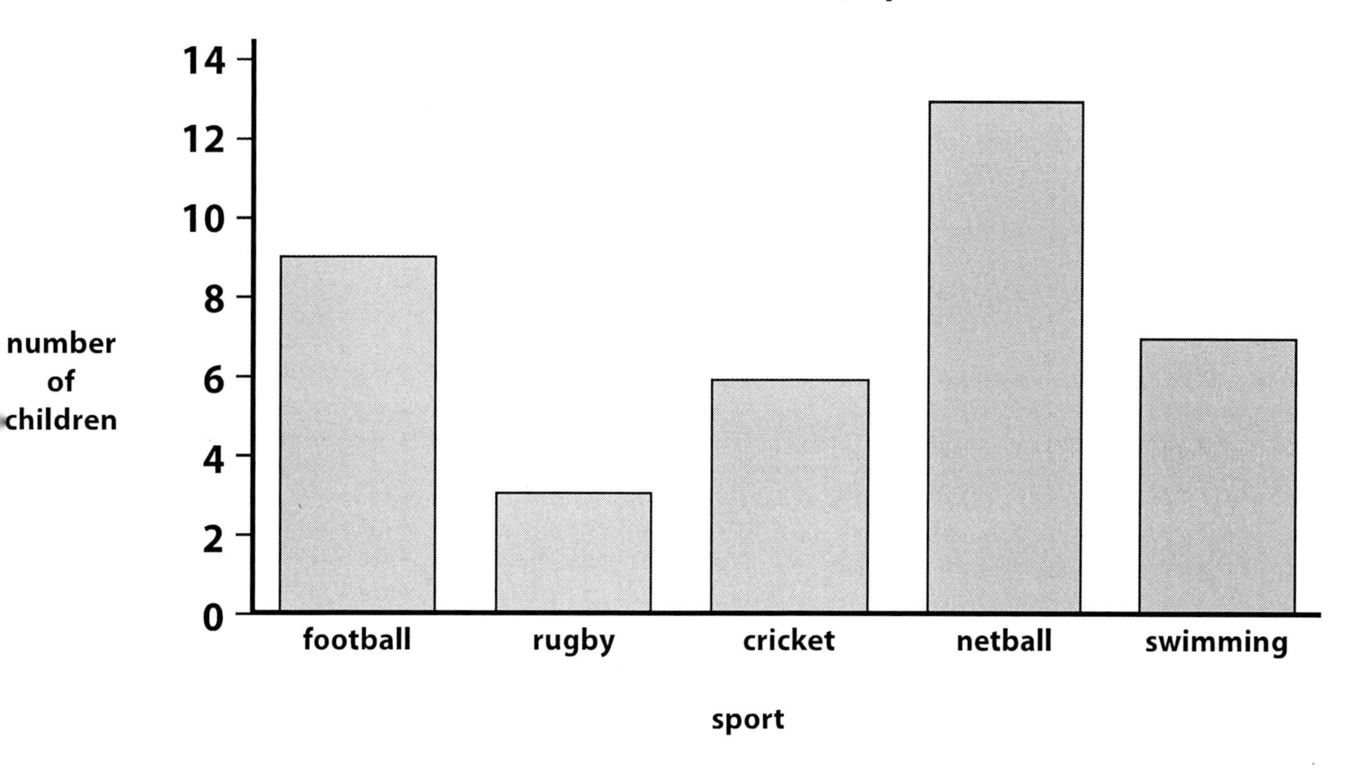

1. What is the most popular sport in Year 3?

2. How many people said they like cricket?

3. How many people said they like swimming?

4. How many people said they like either football or cricket?

5. How many more people like netball than rugby?

6. If **3** people changed their mind and said they liked cricket and not football, how many more people would like cricket than football?

Name ....

Interpret data and solve problems from a bar chart

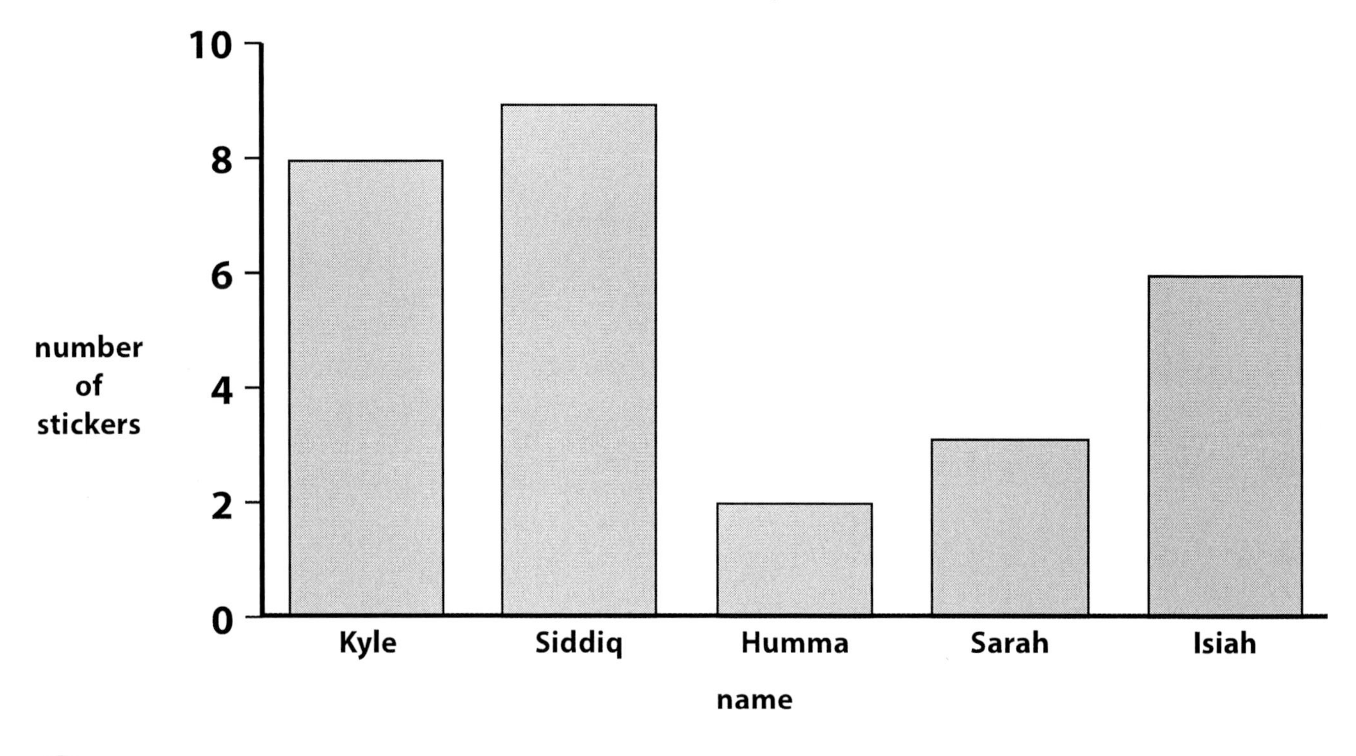

1. Who has collected the most stickers so far?

2. How many stickers have Humma and Sarah collected altogether?

3. How many more stickers has Siddiq collected than Isiah?

4. How many stickers fewer than Kyle has Sarah collected?

5. How many stickers have they collected altogther?

6. If **10** stickers are needed to complete each person's sticker book, how many more stickers do the children need to collect altogether?

Name ....

Interpret data and solve problems from a bar chart

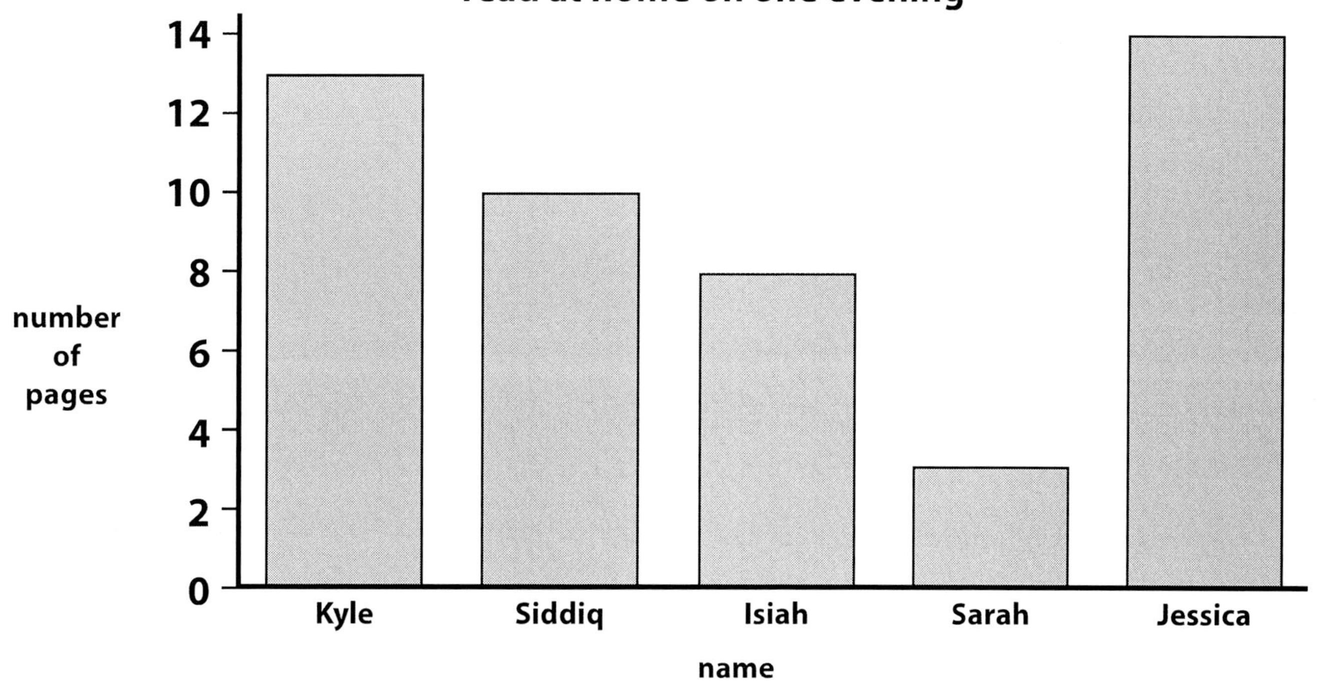

1. Who read the most pages?

2. How many pages did Isiah read?

3. How many pages did Kyle read?

4. How many more pages did Jessica read than Siddiq?

5. Why do you think Sarah read only **3** pages?

   ....

6. How many pages did all **five** children read between them?

Interpret data and solve problems from a pictogram

Pictogram to show how Year 3 travel to school

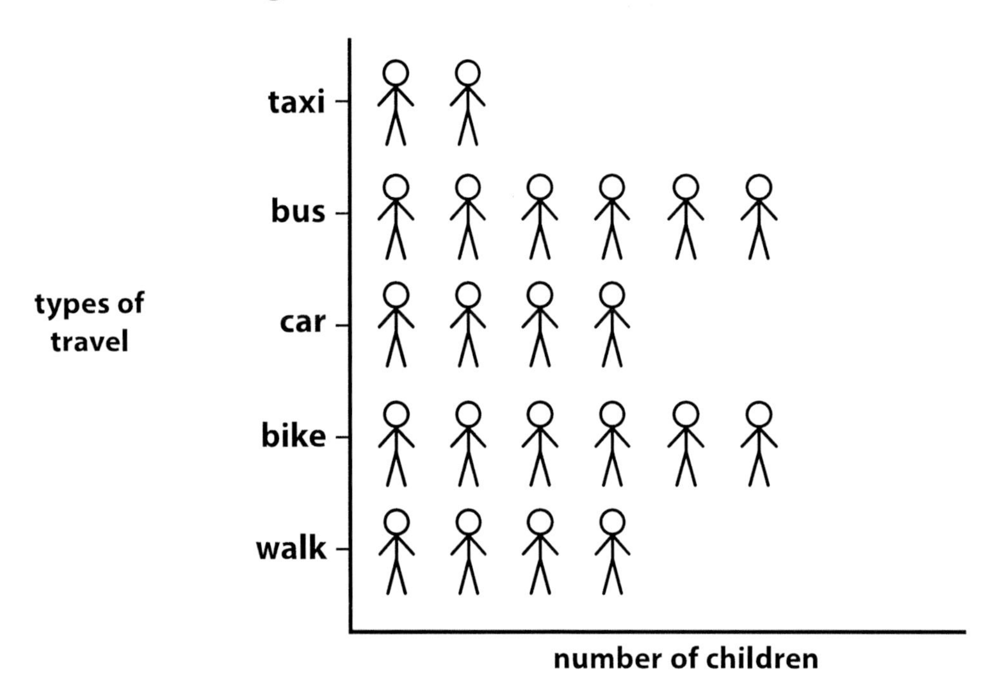

1. How many children go to school by bus?

2. How many children go to school by taxi?

3. How many children go to school either by bike or by walking?

4. How many more children walk than go in the car?

5. How many more children go by bus than by taxi?

6. Do you think that most children live very near to the school? **Yes / No**

   Explain your answer.

   ....

   ....

Name ....

Interpret data and solve problems from a pictogram

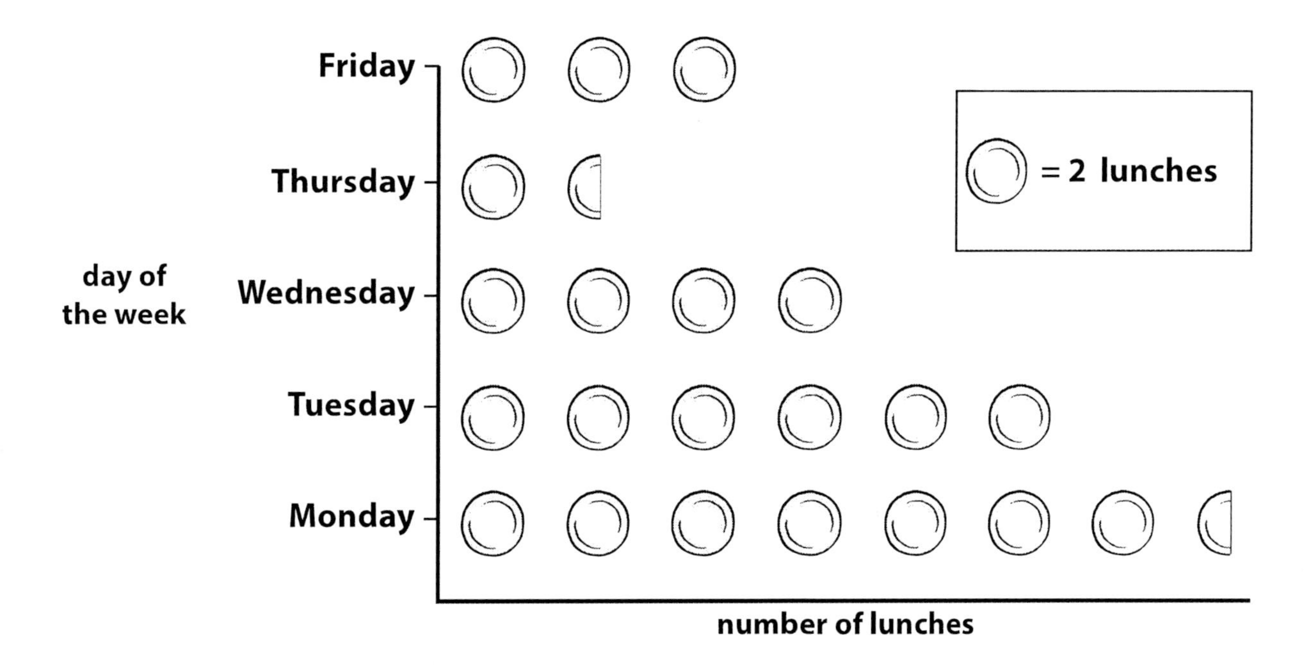

1. How many hot lunches were eaten on Tuesday?

2. What was the most popular day for eating hot lunches?

3. How many more hot lunches were eaten on Tuesday than on Wednesday?

4. How many more hot lunches were eaten on Monday than on Thursday?

5. How many hot lunches were eaten in total during the week?

Name ....

Interpret data and solve problems from a pictogram

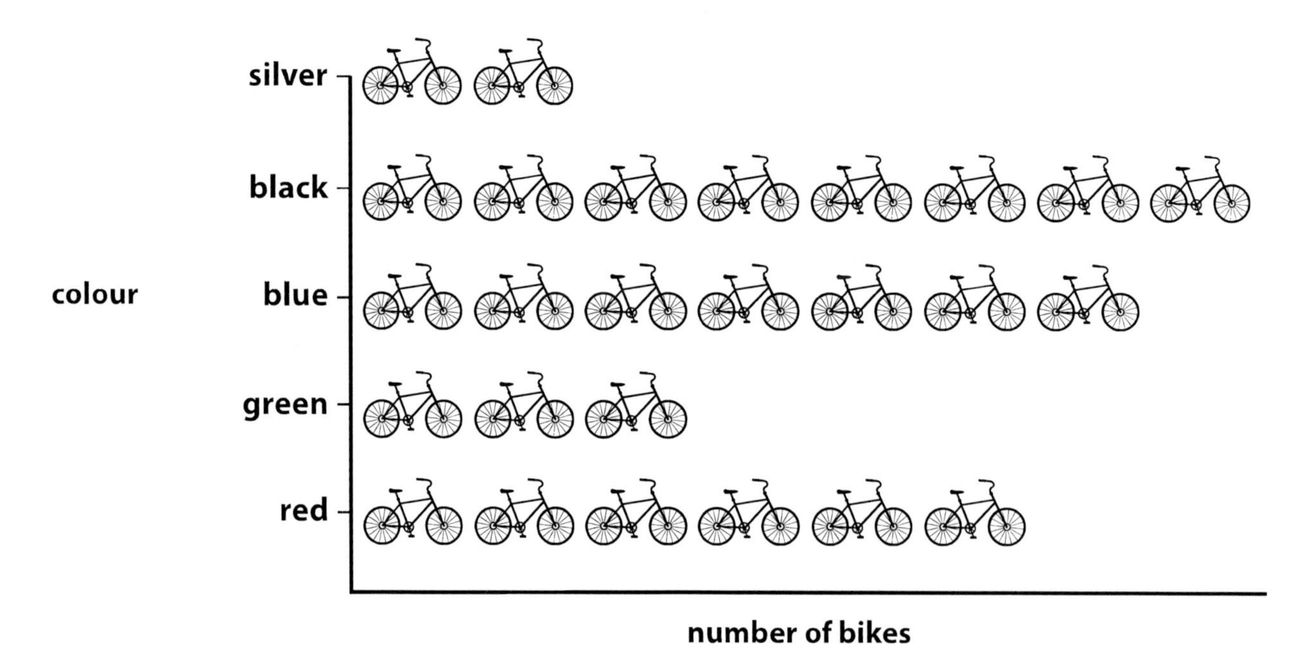

1. What is the most popular colour of bikes sold this week?

2. How many red bikes are sold?

3. How many silver and blue bikes are sold?

4. How many more blue bikes are sold than green bikes?

5. Sarah says, "Black and red together sold **twice** as many as blue." Is she correct?
   Explain your answer. **Yes / No**

   ....

   ....

Name ....

Interpret data and solve problems from a table

Day	Number of cars in the car park
Monday	**11**
Tuesday	**13**
Wednesday	**17**
Thursday	**6**
Friday	**12**

1. How many cars were in the car park on Friday?

2. On which day were there the most cars in the car park?

3. How many cars were there on the **first three** days?

4. How many more cars were there on Wednesday than on Thursday?

5. Which day has **twice** as many cars as the day before?

6. Using the frequency table, write a question that you could ask about cars in the car park.

....

....

Name ....

Interpret data and solve problems from a table

Table to show the number of children in the swimming pool

	10 am	11 am	12 noon	1 pm	2 pm
Saturday	8	7	6	7	12
Sunday	7	8	5	6	11

1. How many children were in the pool on **Saturday** at **10 am**?

2. How many more people were in the pool at **2 pm** on **Sunday** than **12 noon** on **Saturday**?

3. Why do you think there were the least number of children in the pool at **12 noon**?

   ....

   ....

4. How many children were in the pool on **Sunday** afternoon (including **12 noon**)?

5. The lifeguard, said, "There were **half** as many children in the pool at **1 pm** on **Sunday** than there were at **2 pm** on **Saturday**." Is he correct?

   Explain your answer.

   Yes / No

   ....

   ....

Name ....

Interpret data and solve problems from a table

Name of child	Number of team points gained in a week
Sarah	16
Humma	15
Jessica	19
Siddiq	10
Isiah	8

1. Who got the most team points during the week?

2. Who got **twice** as many team points as Isiah?

3. How many team points did Siddiq get?

4. How many more team points did Humma get than Isiah?

5. How many more team points did Jessica get than Isiah and Siddiq put together?

6. How many team points did all **five** children get between them?

Name ....

It may be appropriate for children to use exercise books or paper to record their answers, working out or explanations.

MASTERING

Statistics

A bar chart to show the ice cream flavours chosen by children in Year 3

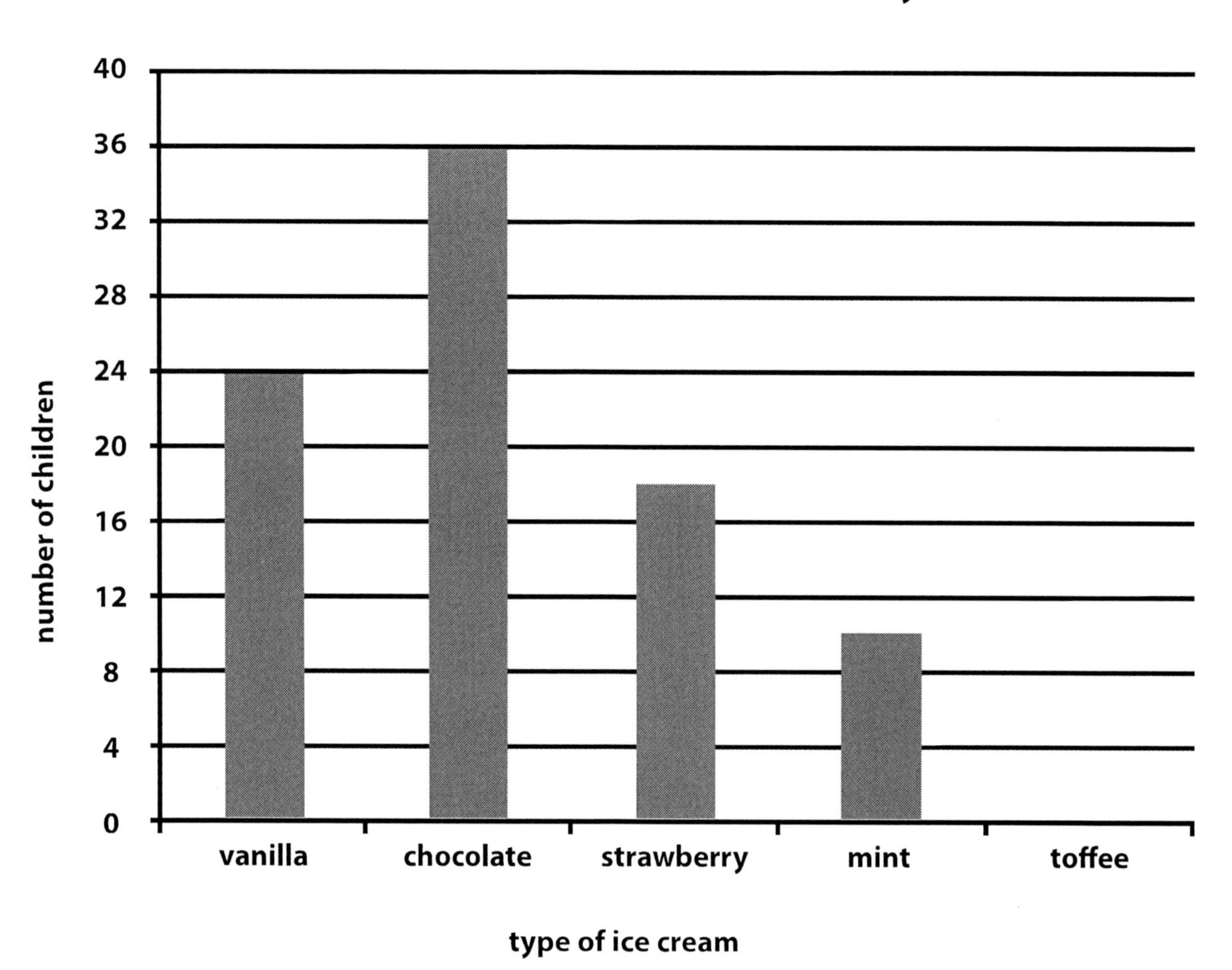

1. There are **120** children in Year 3.
All of them chose an ice cream.
Calculate how many children chose toffee, and then draw a bar on the chart to show this.

2. Siddiq says, "**Twice** as many children chose strawberry as chose mint."
Is he correct? **Yes / No**

Explain your answer.

....

....

Name ....

3 Sarah says, " $\frac{1}{5}$ of the children chose vanilla."

Is she correct? **Yes / No**

Explain your answer.

....

....

4 Which **two** flavours were chosen as many times in total as mint and toffee?

[ ] **and** [ ]

5 Which flavour was twice as popular as strawberry?

[ ]

6 How many children would need to switch from vanilla to chocolate to make chocolate twice as popular as vanilla?

[ ]

7 Make up your own true/false statement about this bar chart.

....

....

Name ....

8 Present the information from the ice cream bar chart in pictogram below.

Use [ice cream symbol] to represent a number of ice creams.

You cannot use it to represent **1** ice cream.

Flavour	Number of ice creams
vanilla	
chocolate	
strawberry	
mint	
toffee	

9 What value does [ice cream symbol] represent in your pictogram?

10 Explain why you chose this value.

....

....

Name ....

Siddiq did a survey of the number of children who played on the climbing equipment in the playground during break time.

Type of equipment	Number of children	Frequency
slide	𝍸 𝍸 𝍸 ||	
climbing frame	𝍸 𝍸 𝍸 𝍸 𝍸 |||	
monkey bars	𝍸 𝍸 ||||	
swings	𝍸 𝍸 |	
balance bar		

The number of children who played on the balance bar was equal to **half** the total of the children who played on all of the rest of the equipment.

Complete the tally for the balance bar.

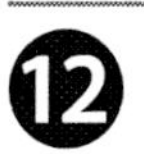

Complete the frequency column to show how many children played on each piece of playground equipment.

Name ....

13 Now use the information to complete this bar chart.
Use a ruler.

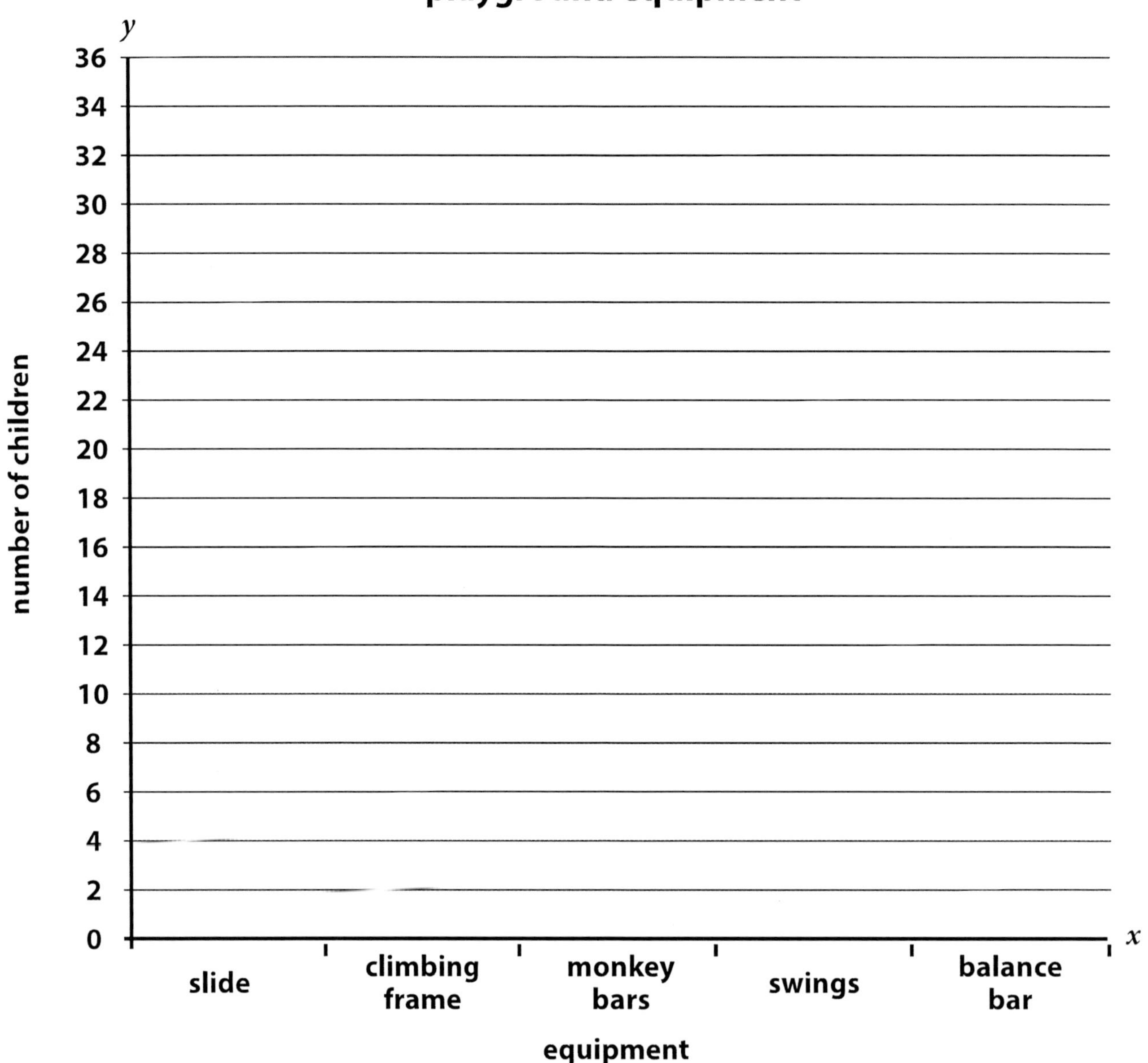

14 The units on the y-axis go up in **2**s.
Why would it not be appropriate to go up in **4**s?

....

....

Name ....

15 Year 3 want to raise money for their charity. They hope to raise **£100** by the end of term. The bar chart shows how much they have raised each week.

How much money have Year 3 raised so far?

£ 

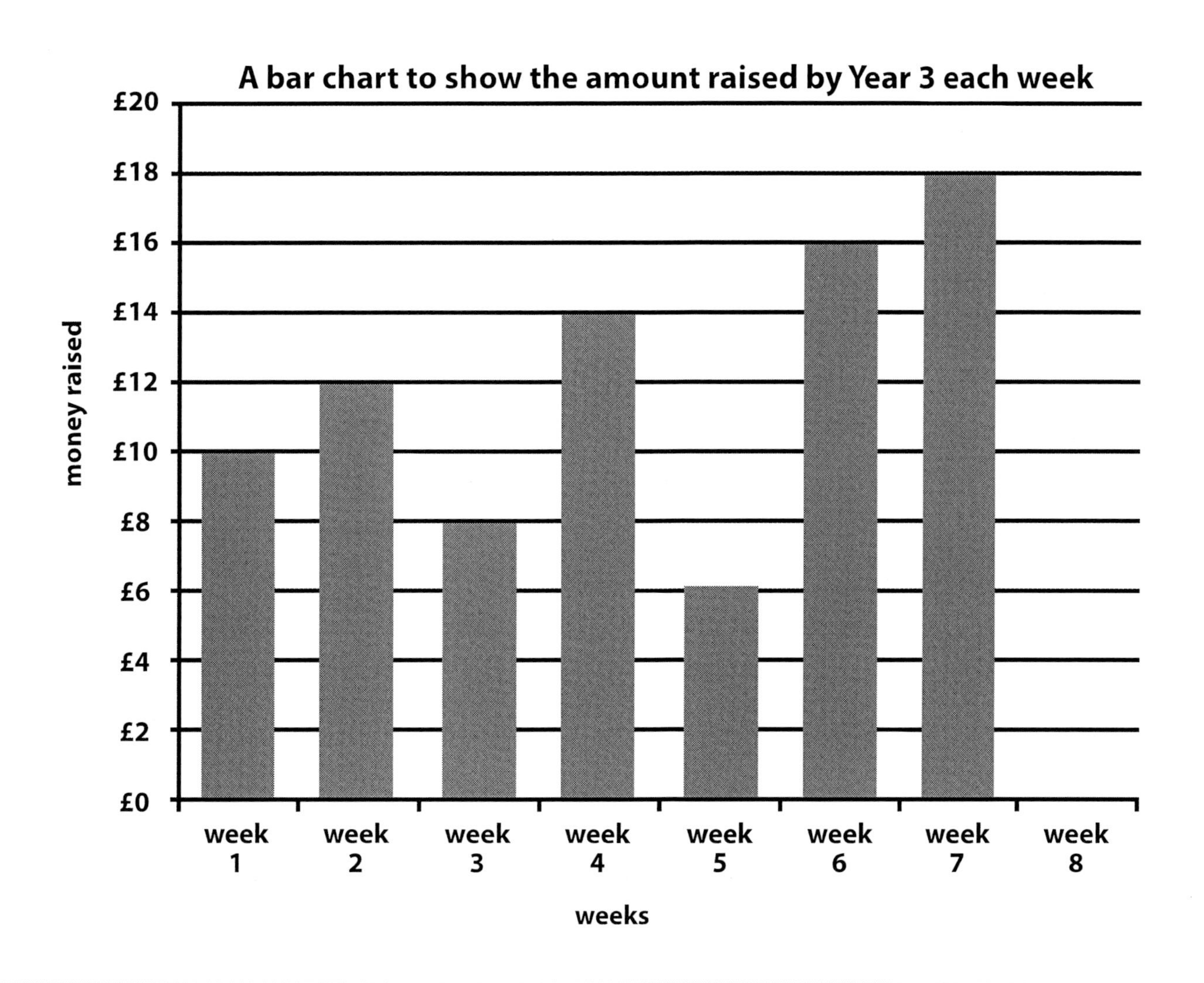

16 **a** How much money do Year 3 need to raise in week 8 in order to reach their target?

£ [ ]

b Show this as a bar on the bar chart. Use a ruler.

Name ....

17 By which week had Year 3 raised exactly **half** of their target amount?

18 In which week did Year 3 raise exactly half the amount they raised in week 6?

19 Use the information presented in the bar chart to draw a pictogram.
Use ⊕ to represent **£4**. Think how you will show **£1**, **£2** and **£3**.

Week	Amount raised
1	
2	
3	
4	
5	
6	
7	
8	

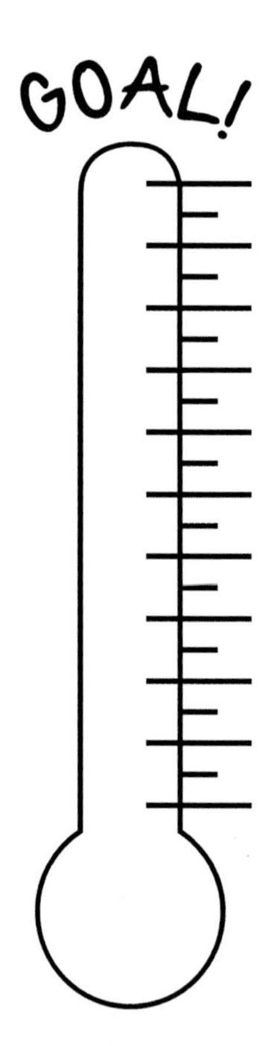

20 Show each week's totals on the Goal thermometer.
You will need to choose a suitable scale.

21 The bar chart and the pictogram show the same information about the number of cars that passed the school last week. Use the information in the pictogram to complete the bar chart.

	Number of cars seen
silver	
black	
red	
blue	
other	

A bar chart to show the colour of cars that passed the school

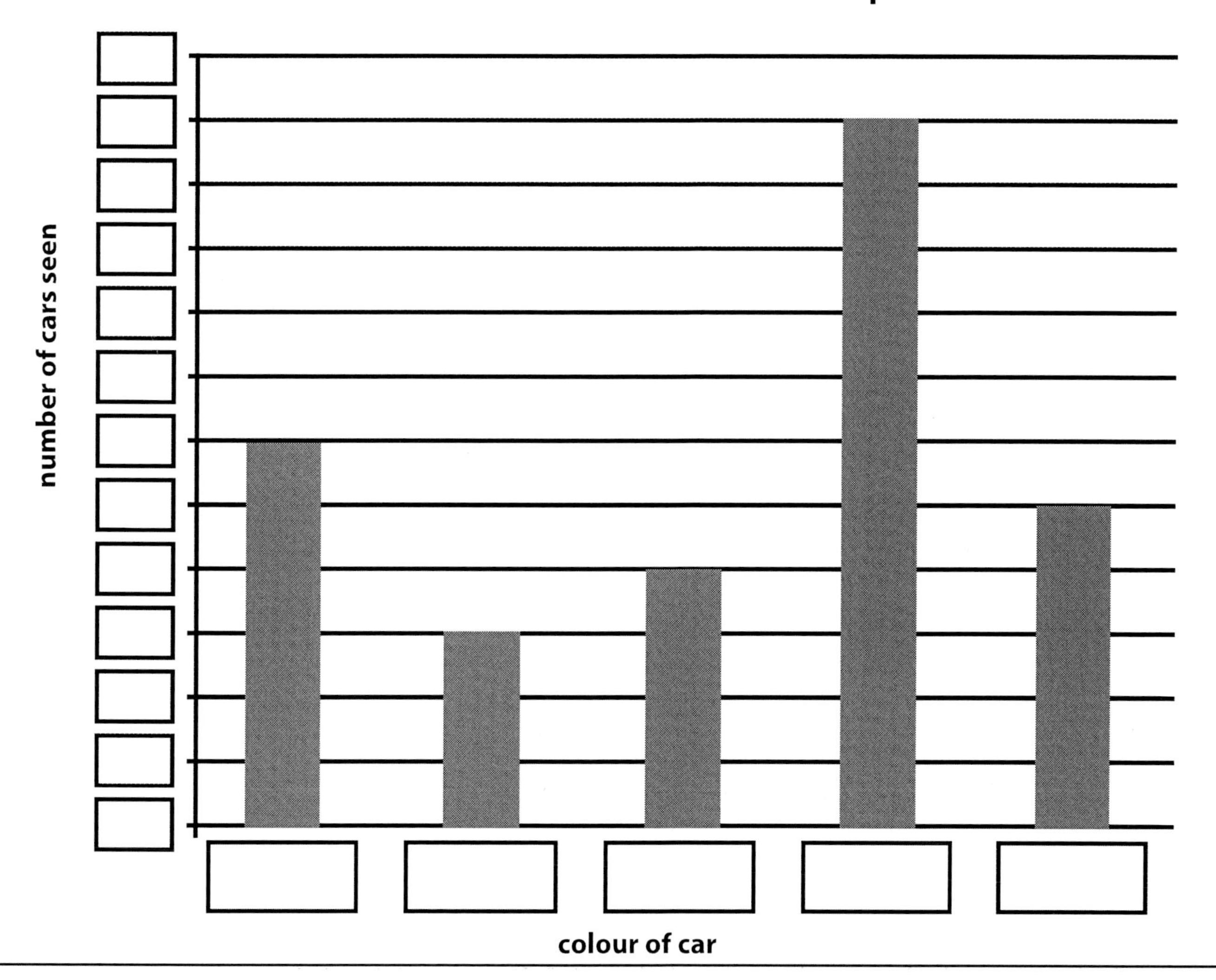

22 How did you work out the interval you used for the y-axis?

....

....

23 Why do you think one bar is labelled 'other'?

....

....

Name ....

It may be appropriate for children to use exercise books or paper to record their answers, working out or explanations.

INVESTIGATION

Trip to Oakham Hall

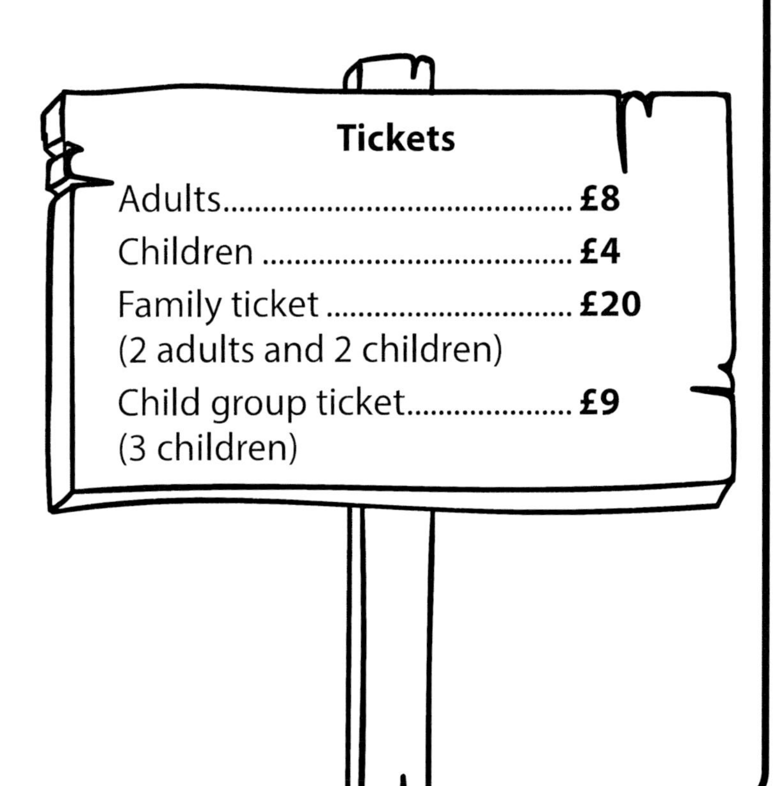

Tickets
Adults.. £8
Children £4
Family ticket £20
(2 adults and 2 children)
Child group ticket..................... £9
(3 children)

Sarah, Kyle and Humma decide to visit the fair being held in the grounds of Oakham Hall.

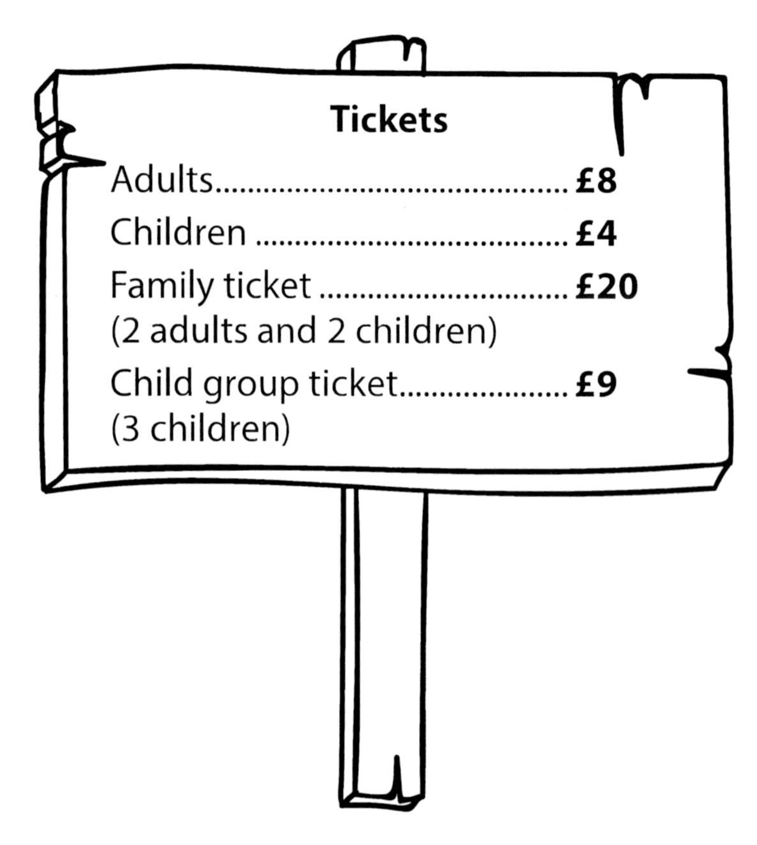

1 How much does it cost in total for the children to go to the fair if they each buy their own ticket?

£ ☐

2 How much do they save if they buy the child group ticket?

£ ☐

Name ....

There is a miniature railway, with one train, which runs between different attractions at the fair.
After the café, it goes straight back to the car park.
Sarah, Kyle and Humma arrive at the entrance at **10.20 am**.

Train Times			
Car park	**10:00**	**10:30**	**11:00**
Entrance	**10:10**	**10:40**	**11:10**
Funfair	**10:15**	**10:45**	**11:15**
House	**10:20**	**10:50**	**11:20**
Café	**10:25**	**10:55**	**11:25**

3 How long do they have to wait until the next train arrives?

minutes

4 At what time would the children arrive at the funfair if they went on this train?

5 It takes **10** minutes to walk to the funfair. The children decide it will be quicker to walk than to wait for the next train.

At what time do the children arrive at the funfair?

6 How many minutes earlier is this than if they had waited for the next train?

minutes

Name ....

The railway runs between the different attractions as shown by the plan below. The railway track between the entrance and the funfair runs vertically on the plan.

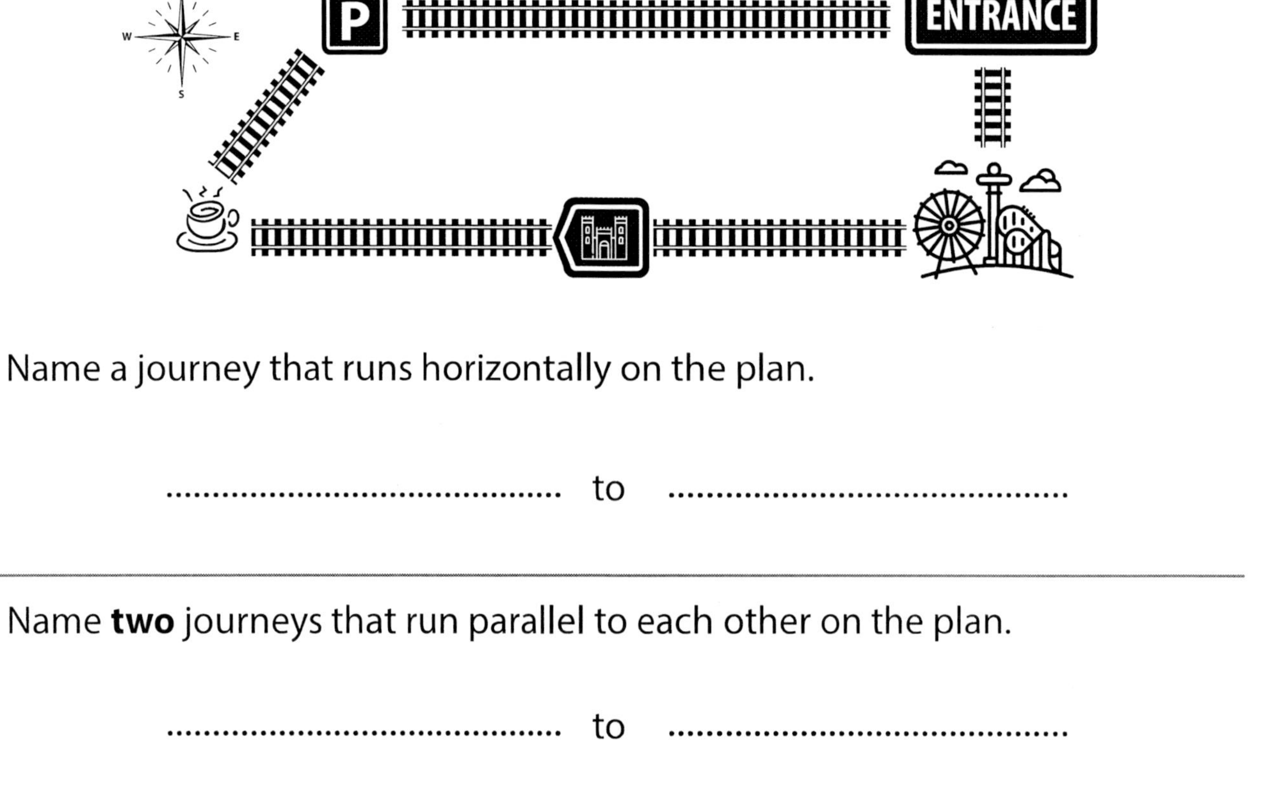

7 Name a journey that runs horizontally on the plan.

.... to ....

8 Name **two** journeys that run parallel to each other on the plan.

.... to ....

.... to ....

9 The entrance is directly east of the visitor parking. Sarah, Kyle and Humma walk from the entrance to the funfair.

In what direction do they walk?

10 From the funfair, the children make a **quarter** turn to look at Oakham Hall.

In which direction do they turn? Circle the correct answer.

clockwise **anti-clockwise**

Name ....

11 Sarah, Kyle and Humma buy 5 raffle tickets each.
Use these clues to find the numbers on Sarah's tickets.

- The numbers on Sarah's tickets are all consecutive **2-digit** multiples of **5**.
- The product of the digits in the largest number is **40**.
- The difference between the digits in the largest number is **3**.

12 Use these clues to find the numbers on Kyle's tickets.

- The numbers on Kyle's tickets are all consecutive **2-digit** multiples of **8**.
- The largest number is between **50** and **100**.
- The sum of the digits of the largest number is **9**.
- The difference between the digits is **5**.

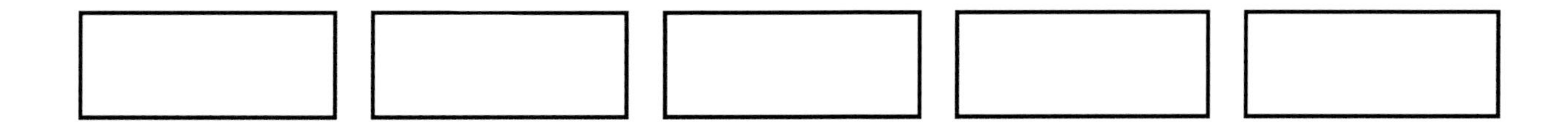

13 Use these clues to find the numbers on Humma's tickets.

- The numbers on Humma's tickets are all consecutive **2-digit** numbers.
- The total of the numbers is **110**.

Name ....

14 Find the total of the numbers on Sarah's and Kyle's tickets. Use the grid below for your calculations:

Sarah's tickets ☐ **Kyle's tickets** ☐

---

15 At the fair, Sarah has a go with the Hoopla.
She has **5** bean bags. She manages to throw **one** bean bag into each of the **five** hoops. Unfortunately, the signs have fallen off the hoops.
She does know that the values of the hoops are the first **five** multiples of the **4** times table.

Use these clues to find the value for each hoop.
Write the values in the hoops.

- The first **three** hoops add up to **40**.
- The last **three** hoops add up to **32**.
- The value of the last hoop is double the value of the first hoop.

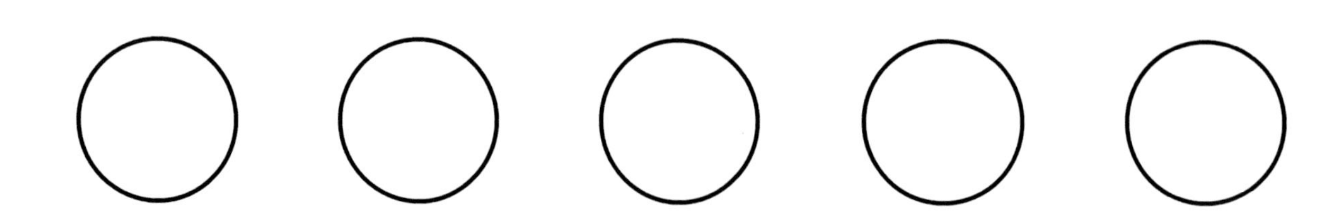

Name ....

Kyle decides to have a go at the Hoopla.
The hoops are kept in the same order as for Sarah's go.
Kyle manages to throw all **five** bean bags into the hoops, but does not get **one** into each hoop.

Use these clues to find which hoops Kyle threw his bean bags into.
Label the hoops, then write the number of bean bags inside the hoops.

- The total of the **5** bean bags was **56**.
- Kyle did not throw a bean bag into either the largest or the smallest hoops.

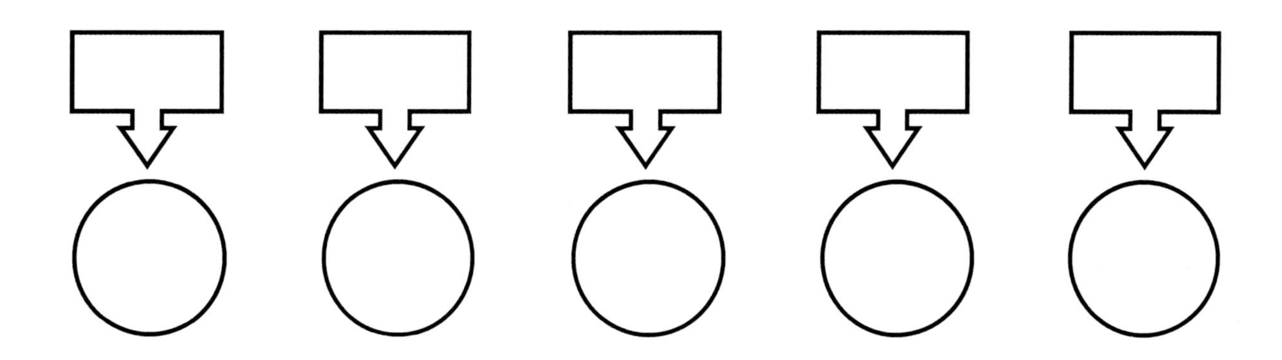

Humma had a go at the Hoopla. She scored **20** points.

List the different ways that Humma could have scored **20** points.
She does not need to score with all **five** bean bags.
For example, she could have scored **8 + 12**.

8 + 12 = 20	

Name ....

The children decide to try Splat a Rat.

18 What **3D** shape is the tube?

19 Splat a Rat costs **50p** for **three** splats.
Sarah spends **£1**.

How many splats does Sarah have?

20 Kyle spends **£1.50**.

How many splats does he have?

21 Humma spends **50p** and splats **2** rats.

In what fraction of her splats did Humma splat a rat?

22 In total, the children managed to splat a rat in **9** of their splats.

What fraction of the children's splats were successful overall?

Name ....

The children take turns at Ring the Bell.
They can score **1**, **2**, **3**, **4** or **5** points.

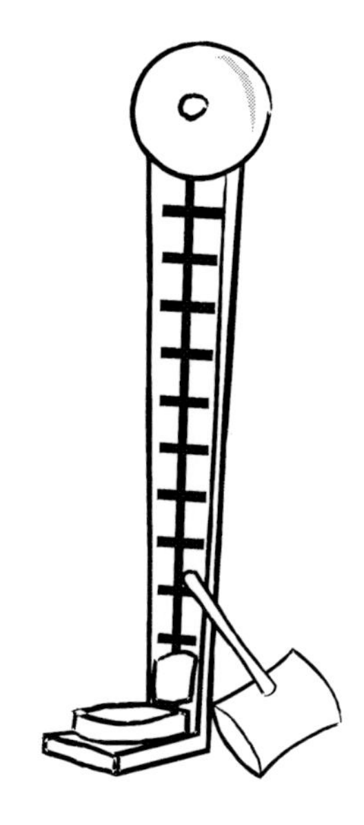

- Humma scores the highest.
- Sarah's score is an even number.
- Each child's score is different.
- The total of their scores is **12** points.

How many did each child score?

Sarah ______ **points**

Kyle ______ **points**

Humma ______ **points**

On their second go, the children scored a total of **14** points.
Humma's score was the lowest.

How many did each of the children score in their second go?

Sarah ______ **points**

Kyle ______ **points**

Humma ______ **points**

Name ....

The children decide to have a go on the Dodgems.
Each go on the Dodgems lasts **5** minutes.
There are **16** Dodgem cars.
They are **35th**, **36th** and **37th** in the queue for the Dodgems.

The dodgems have just finished.

How long do the children have to wait for their go?

minutes

26 The children have **3** goes on the Dodgems. They have to wait for **one** ride in between each go.

How much time in total do the children spend on the Dodgems? (Don't include the wait at the beginning.)

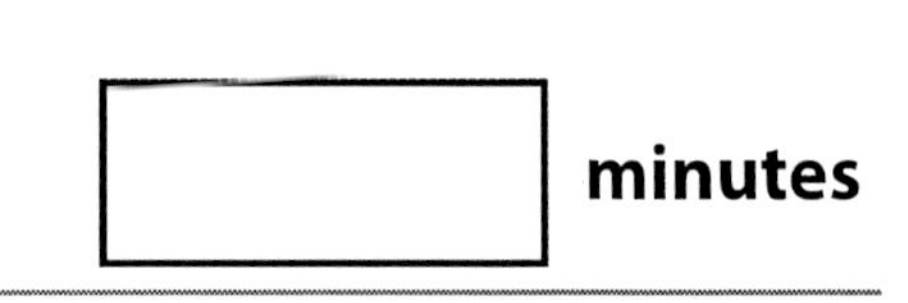

27 The children start their first ride at **12.15 pm**.

At what time do the children finish their Dodgem rides?

Name ....

As soon as they have finished on the Dodgems, the children decide they are hungry.
The café is a **20 minute** walk away from the Funfair.

They look at the train timetable.

Train Times			
Car Park	**12:00**	**12:30**	**13:00**
Entrance	**12:10**	**12:40**	**13:10**
Funfair	**12:15**	**12:45**	**13:15**
House	**12:20**	**12:50**	**13:20**
Café	**12:25**	**12:55**	**13:25**

Will it be quicker for the children to walk or to wait for the next train?

walk / train

Explain your answer.

....

....

The train can hold **32** passengers. There are **4** carriages.

How many passengers can fit in each carriage?

Name ....

When the children travel from the funfair to the café, there are a total of **16** passengers on the train. There are an odd number of passengers in each carriage.

How many passengers could there be in each carriage?
Train 1 has been done for you.

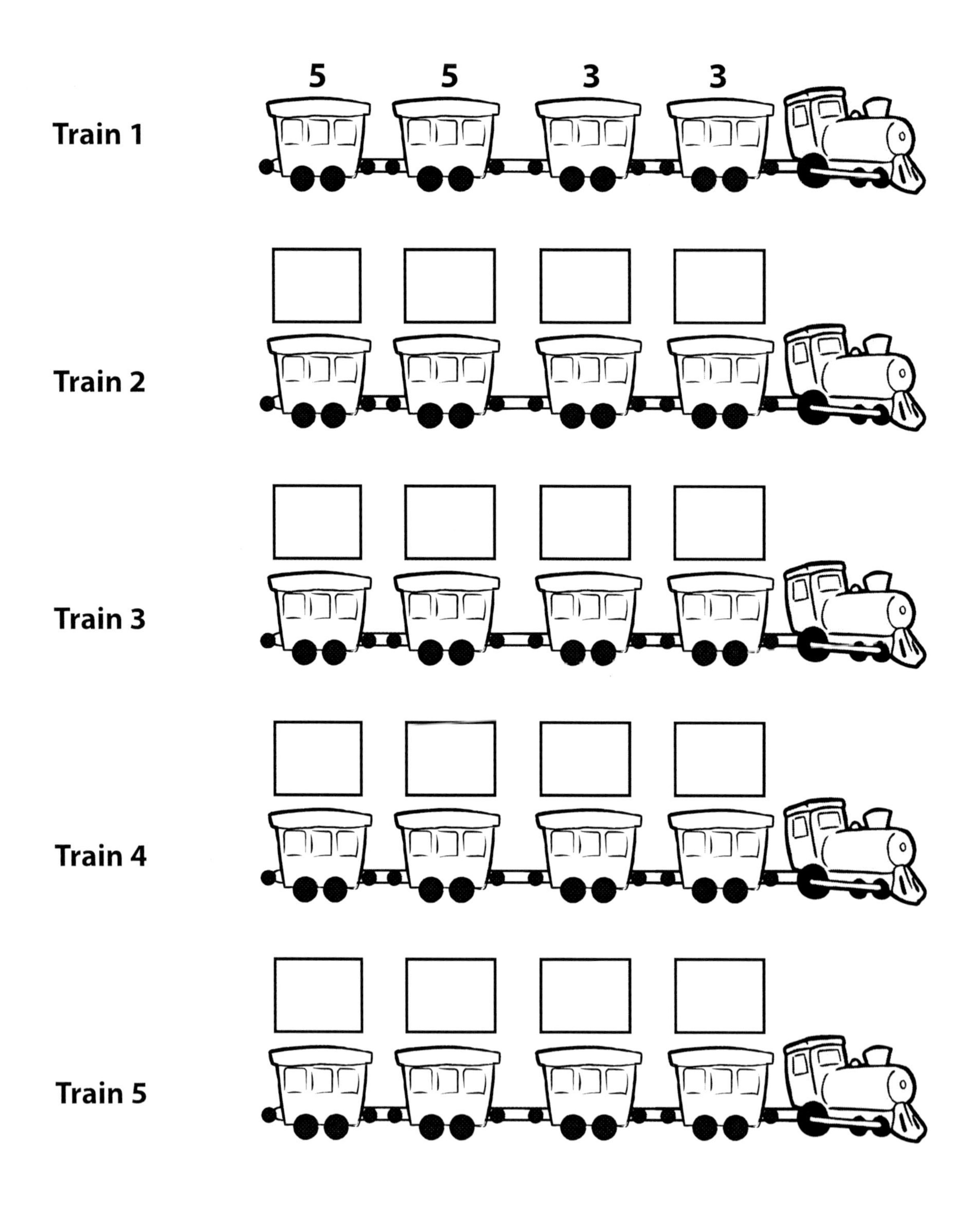

Name ....

The train that the children catch has a different number of passengers in each carriage.
Which number train is this?

When the children get off the train at the café, some other passengers also get off and others get on. The train is still not full.

- When Sarah divides the number of passengers by **3**, there is a remainder of **2**.
- When Kyle divides the number of passengers by **4**, there is a remainder of **2**.
- When Humma divides the number of passengers by **5**, there is a remainder of **4**.

How many passengers are left on the train? 

Name ....

33 The children go to the café to have lunch.
All the chairs have **4** legs and the tables have **3** legs.
Altogether, the children can see **60** legs.
There is an even number of tables and an even number of chairs.
There are fewer tables than chairs.
The number of chairs is a multiple of **12**.

How many chairs and tables are there in the café?

☐ **chairs** ☐ **tables**

---

34 The chairs are shared equally between the tables in the café.

How many chairs are around each table?

☐

---

35 The tables have square tops with sides measuring **75 cm**.

What is the perimeter of the top of each table?

☐ **cm**

Name ....

 The clock on the wall shows this time:

What time does the clock show?

 The children are at the café and they have to be at the entrance by **2.30 pm**. The train travels in a loop and it takes **15 minutes** to get to the entrance from the café.

Train Times			
Car Park	**13:30**	**14:00**	**14:30**
Entrance	**13:40**	**14:10**	**14:40**
Funfair	**13:45**	**14:15**	**14:45**
House	**13:50**	**14:20**	**14:50**
Café	**13:55**	**14:25**	**14:55**

What time is the last train the children could catch from the café in order to get to the entrance by **2.30 pm**?

 Draw the hands on this clock to show the time the children need to leave the café if they are going to walk back to the entrance by **2.30 pm**.

Name ....

In the café, the children can choose one main, one type of potato and one side.

Main	Potato	Side
Sausages (S)	**Fries (F)**	**Baked Beans (BB)**
Fish Fingers (FF)	**Mash (M)**	**Peas (P)**
Chicken Nuggets (CN)		

List all the combinations that they could choose.
One has been done for you.

S + F + BB		

40 The prices for the meals are displayed on a board in the café.

How much does the most expensive meal cost?

Sausages	**£1.45**
Fish Fingers	**£1.25**
Chicken Nuggets	**£1.55**
Fries	**55p**
Mash	**50p**
Baked Beans	**35p**
Peas	**40p**

£ ☐

41 Kyle bought the cheapest meal.
How much did he pay?

Sausages	**£1.45**
Fish Fingers	**£1.25**
Chicken Nuggets	**£1.55**
Fries	**55p**
Mash	**50p**
Baked Beans	**35p**
Peas	**40p**

£ [ ]

42 What is the difference in price between the most expensive and the cheapest meals?

[ ] **p**

43 Sarah wants to buy sausages, mash and peas. How much change does she get from **£5**?

£ [ ]

44 Humma spent **£2.40**. Which **three** items did she buy?

[ ] + [ ] + [ ]

Name ....

The café owner makes a record of the sales every day. The bar charts show the sales of sausages, fish fingers and chicken nuggets over the last **two** weeks.

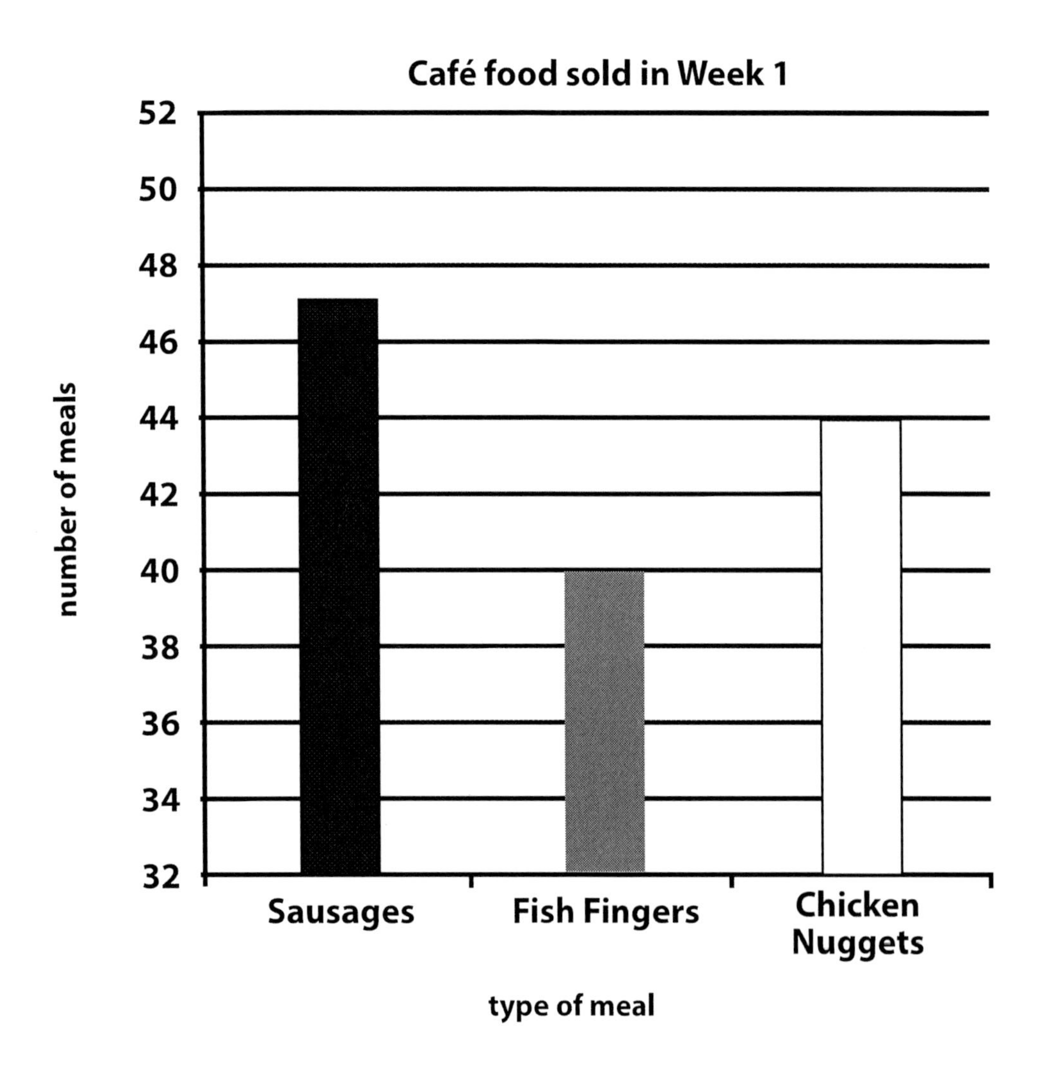

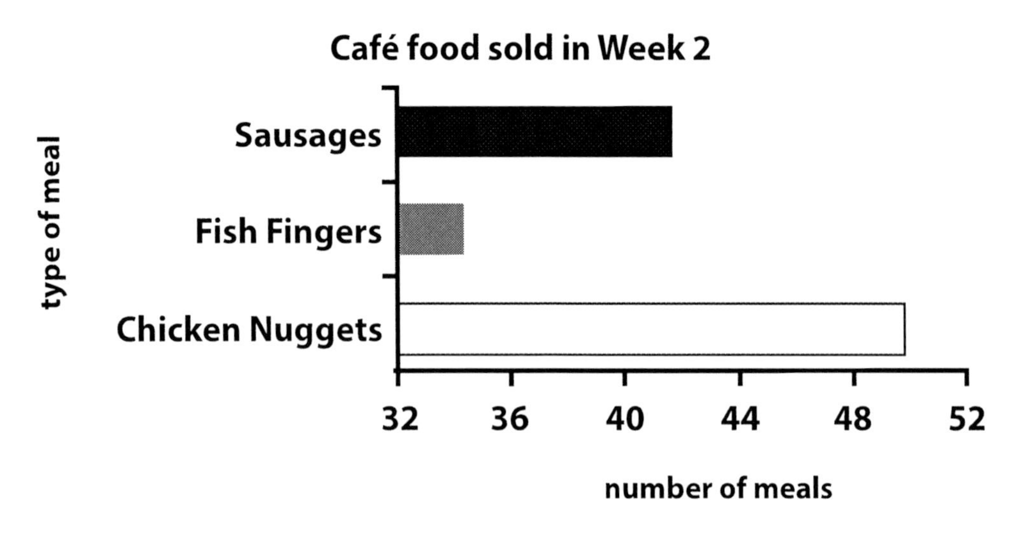

Name ....

45 How many meals were sold in:

Week 1 ☐ **meals**

Week 2 ☐ **meals**

46 How many of each type of food was sold altogether?

☐ **Sausages** ☐ **Fish Fingers** ☐ **Chicken Nuggets**

47 Write the names of the food in order from most popular to least popular.

☐ ☐ ☐

48 How many more chicken nuggets were sold than fish fingers?

☐

49 Choose your own meal from the café and find how much it would cost.

☐ + ☐ + ☐

Total cost **£** ☐

Name ....

Year 3: NUMBER – Number and place value

Page 1: 1) 8 2) 12 3) 16 4) 32 5) 50, 100 6) 300

Page 2: 1) 12 2) 8 3) 300 4) 50, 100, 150 5) 400 6) 12, 20, 24

Page 3: 1) 4 2) 4 3) 12 4) 9 5) 866 6) 5

Page 4: 1) 9 2) 12, 14, 16 3) 30 4) 20, 24 5) 190 6) 232, 332, 432, 532

Page 5: 1) 100 2) 77 3) 74 4) 15 5) 324 6) 17

Page 6: 1) 48 2) 247 3) 76 4) 82 5) 434 6) £156

Page 7: 1) 2 2) 3 3) 5 4) 4 5) 60 6) 329

Page 8: 1) 400 + 80 + 2 2) 287 3) 746 4) no; appropriate explanation 5) no; appropriate explanation

Page 9: 1) 428; appropriate explanation 2) 320, 310, 305, 300 3) 536 4) 349 5) 345, 354, 435, 534, 543

Page 10: 1) 820, 840, 865, 882 2) 895; appropriate explanation 3) 690, 658, 641, 630 4) 490 5) 981, 918, 899, 891, 819

Page 11: 1) 30 2) Humma (7 m) 3) 10 m 4) girls (36) 5) 50 6) Humma (135 cm)

Page 12: 1) 100 ml 2) yes; appropriate explanation 3) 800; appropriate explanation 4) 5 5) 700 ml

Page 13: 1) seventy two 2) 230 3) 343 4) 862 5) seven hundred and eighty seven 6) eight hundred and fifty three

Page 14: 1) eight 2) 15 3) two hundred and fifty two 4) 464 5) three hundred and forty eight 6) eight hundred and seventy six

Page 15: 1) 90p 2) £2.67 3) £100 4) £398 5) 5 in £650 6) £126

Page 16: 1) 90 ml 2) Wednesday (8 lengths) 3) Isiah (132 cm) 4) 10 kg 5) 267 miles circled

Page 17: 1) 20 2) 30 3) 46 4) 204 5) eldest - Joe, youngest - Henry 6) appropriate explanation

Page 18: 1) 112 2) two hundred and fifty six 3) 900 g 4) 643 5) 16

Page 19: 1) 50 2) 452 3) 9 4) 4 5) appropriate examples 6) 32

Page 20: 1) 24 2) one hundred and forty three 3) 126 4) 260 g 5) appropriate explanation 6) 80

MASTERING - Number and place value

Page 21: 1 a) 400, 220, 31,13 b) 4, 22, 40, 103, 121, 130, 112, 202, 211, 301, 310 c) 310; it has the biggest number of hundreds and tens possible d) 4; it has the smallest place value column

Page 22: 2) 153 is missing 3) Isiah 557, Siddiq 767, Kyle 75, Sarah 757

Page 23: 4) false, true, false 5 a) 200 cm b) 600 cm

Page 24: 6) 8 hundreds + 3 tens + 7 ones, 83 tens + 7 ones, 8 hundreds + 37 ones
7 a) multiples of 8 - 16, 32, 24, multiples of 5 - 10, 15, 20, 25, multiples of both - 40
b) 23 would be placed outside the diagram c) 80

Page 25: 8) 843 837, 841 817, 842 827 9 a) example 400, 500, 600, 700, 800, 900 b) any appropriate answer

Page 26: 10 a) 536 b) five more counters added to the 10s column c) five hundred and eighty six 11 a) true; appropriate explanation b) 5, 1, 3

Page 27: 12 a) 405 b) 338 13) 452 (451 - 500), 198 (151 - 200), 349 (301 - 400), 255 (251 - 300), 423 (401 - 4 50), 379 (351 - 400)

Page 28: 14 a) 632 b) 696 c) 107 15) set 1) 114, 186, 414, 571 set 2) 124, 286, 424, 572 set 3) 134, 386, 434, 573

Page 29: 16 a) 57 or 75 b) 794, 776, 758 17) 473, 464, 455

Page 30: 18) 532 - look at the value of the hundreds column first and if there is more than one number with the same value then look at the tens column value 19 a) 842 b) 142

Page 31: 20) 426 - tens, 215 - hundreds 392 - ones 21) Isiah 123, Humma 246, Kyle 513

Page 32: 22) any of 124, 133, 142, 151, 160, 241, 313, 412, 511, 610, 421, 331, 214, 115, 106, 223, 232, 322, 304, 205, 403, 502, 601, largest possible 610, smallest possible 106
23) 250, 400, 800, 500, 400, 800, 700, 600, 550

Page 33: 24 a) C b) A and C c) any answer between 500 -700

Year 3: NUMBER - Addition and subtraction

Page 34: 1) 119 2) 259 3) 112 4) 134 5) 146 6) 139

Page 35: 1) 115 2) 120 3) 166 4) 138 5) 232 miles 6) 126

Page 36: 1) 111 2) 225 3) 132 4) 393 5) £86

Page 37: 1) 142 2) 144 3) 210 4) 47 5) 243

Page 38: 1) 40 2) 92 3) 88 4) 122 5) 34 6) 152

Page 39: 1) 12 2) 110 3) 150 cm 4) 113 5) 46 cm 6) 213

Page 40: 1) 208 2) 325 3) 337 4) 638 5) 447 6) 878

Page 41: 1) 150 2) 80 3) 157 4) 80 5) 317

Page 42: 1) 345 2) 128 3) 475 g 4) 362 5) 356

Page 43: 1) 88, 130 2) 142 3) 381 4) 330

Page 44: 1) £34 2) 64 3) 48 4) 348 5) 339

Page 45: 1) 213 2) 231 3) 440 4) 157 5) 380 cm

Page 46: 1) yes; appropriate explanation 2) 50 + 50 3) yes ; appropriate explanation 4) 100 5) 100 6) 450

Page 47: 1) 10 – 2 = 8 or 10 – 8 = 2 2) no; appropriate explanation 3) 7 + 8 = 15 4) 16 – 7 = 9 or 16 – 9 = 7 5) 112 – 80 = 32 or 112 – 32 = 80 6) 19 + 78 = 97 or 78 + 19 = 97

Page 48: 1) appropriate explanation 2) 36 3) 16 4) 698 5) 20 – 8 = 12, 21 – 8 = 13, appropriate calculation

Page 49: 1) 97p 2) £3 3) £5 4) £2.49 5) 35p 6) no; appropriate explanation

Page 50: 1) 42p 2) £2.25 3) £4.75 4) £72 5) £63 6) £60

Page 51: 1) 90 2) 37p 3) 49 4) 36 5) 116 6) 215

Page 52: 1) 45 2) 41p 3) 158 4) 30 5) 86

Page 53: 1) 87 2) £17 3) 196 4) 20 5) 616 6) 356

MASTERING - Addition and subtraction

Page 54: 1) a) 862 b) 299 c) 693 d) 297 2) a) £3.70 b) 18p c) £3.52

Page 55: 3) 224, 256 + 224 = 480, 480 – 224 = 256, 224 + 256 = 480 4) always, never, always - any appropriate examples

Page 56: 5 a) yes; appropriate explanation b) yes; appropriate explanation 6) 27p, no; it should be worked out by subtracting 35p from 62p

Page 57: 7) any appropriate combinations, appropriate explanation 8)179, Sarah is correct, to find 'more than' the amounts need to be added together

Page 58: 9) appropriate number combinations 10) £27

Page 59: 11) 77 + 22 + 72 12) 586 + 257 = 843 13) 24 + 76 = 100, 56 – 29 = 27, 80 – 35 = 45, 46 + 28 = 74, 500 + 350 = 850

Page 60: 14) no; it depends on the value of the tens 15) 99 – 36, 98 – 35, 97 – 34, 96 – 33, 95 – 32, 94 – 31, you need to work systematically in order to make sure you have all the answers

Page 61: 16) yes; appropriate explanation 17) 70, 30, 60, 40, 50 18) 30, 146, 176

Page 62: 19) a) £1.14 b) 74p c) 2 pencils and 1 sharpener

Page 63: 20) answers given from the top of the pyramid down a) 299, 150 149, 65, 84 b) 352, 184, 168, 87, 81, 49 c) 260, 90, 140, 70, 50

Page 64: 21 a) 165 b) 62 c) 153 d) hockey e) football

Page 65: 22 a) 21 b)16 c) Siddiq 15 Isiah 24 Kyle 23 23) 4, 4

Page 66: 24 a) £1.15 b) yes; chicken salad, ice cream, water

Year 3: NUMBER - Multiplication and division

Page 67: 1) £12 2) 15 3) 12 4) 21 miles 5) 30 6) £2.40

Page 68: 1) 8 2) 12 3) 16 4) 40 5) 28 6) 48

Page 69: 1) 16 2) 24 3) 48 4) 56 5) 64 6) 96

Page 70: 1) 16 2) 12 3) 24 4) 64 5) 44 6) 36

Page 71: 1) 8 2) 32 3) 30 4) 24 5) 96 6) 49

Page 72: 1) 3 2) 6 3) 5 4) 8 5) 9 6) 12

Page 73: 1) 2 2) 3 3) 4 4) 6 5) 8 6) 12, 13

Page 74: 1) 2 2) 4 3) 7 4) 8 5) 8 6) 11, 96

Page 75: 1) 3 2) 2 3) 6 4) 3 5) 12 6) 9

Page 76: 1) 3 2) 4 3) £7 4) 12 5) 9 6) 12

Page 77: 1) 12 2) 5 3) 8 4) 9 5) £8 6) 13

Page 78: 1) 24 2) £40 3) 48 4) 7 5) £36 6) 6

Page 79: 1) appropriate explanation 2) appropriate explanation 3) 16 4) 64; appropriate explanation 5) 48; appropriate explanation

Page 80: 1) 48 2) 80 m 3) £1.20 4) 69 5) 132 6) 212

Page 81: 1) 96 2) 76 3) 150 4) £104 5) 135 6) 168

Page 82: 1) 9 2) 8 3) 20p 4) 15 5) 40 6) £13

Page 83: 1) 14 2) 8 3) 15 4) 32 5) 12 6) 21

Page 84: 1) 88 2) 72 3) 260 4) 136 5) 128p 6) 24 x 8

Page 85: 1) 21 2) 14 3) 24 4) 12 5) 12, 3

Page 86: 1) £64 2) 13 3) 11, 6 4) 78 5) 368 6) 32, 1

Page 87: 1) 400 cm 2) 56 3) 2 m 4) 90 g 5) 104

MASTERING - Multiplication and division

Page 88: 1) 4 x 8 = 32, 32 ÷ 4 = 8, 32 ÷ 8 = 4 2) digits going across the table = 8, 4, 5, 2, 3 digits going down the table = 4, 3, 5, 2, 8 rows from left to right = 20, 8, 12, 24, 12, 6, 9, 40, 20, 25, 15, 16, 8, 10, 6, 64, 32, 40, 16 3) 70p

Page 89: 4) 5 x 6 = 30 10 x 6 = 60 15 x 6 = 30 + 60 = 90 5) 1 x 24, 2 x 12, 3 x 8, 4 x 6 6) 72 ÷ 3 = 24, 12 x 4 = 48, 688 ÷ 8 = 86, 76 x 5 = 380

Page 90: 7 a) 60p b) £7.20 c) £23.40 8) 24, three into each pot

Page 91: 9 a) 3 biscuits each with 2 left over b) 9 children c) 35

Page 92: 10) 5 x 8 ÷ 2 = 20, 2 x 3 x 5 = 30, 3 = 24 ÷ 8 11) =, <, >, <, =

Page 93: 12) no; show multiples of 3 that are even 13) 2 x 15, 3 x 10, 5 x 6, 30 x 1
14 a) 27 x 5 = 135 b) 43 x 5 = 215 c) 35 x 4 = 140 or 85 x 4 = 340 d) 276 x 2 = 552

Page 94: 15 a) 8 b) 6 c) 45 d) 40 e) 9 f) 1 16 a) 2 x 45 = 90, 2 x 54 = 108, 4 x 25 = 100, 4 x 52, 5 x 24, 5 x 42 b) 5 x 42 c) 2 x 45

Page 95: 17) boys; because 16 ÷ 4 = 4 and 15 ÷ 5 = 3 18) 70 cm 19) 25 m 20) 27, 45, 54; multiples of 9

Page 96: 21) yes; show using the 4 and 8 times table 22) 8 tables and 32 chairs 23) 4

Page 97: 24) no; the answer could be a 3-digit number 25) 320, 320, 160, 192 26) 7

Page 98: 27 a) 32 x 5 = 160 b) 25 x 3 = 75, 23 x 5 = 115 28) 36

Page 99: 29) 6 x 8 = 48, 12 x 4 = 48, 1 x 48 = 48, 3 x 16 = 48, 2 x 24 = 48 30 a) larger b) larger c) larger d) smaller

Page 100: 31) 10 32) 256 33) 4 34) x 20 4, 8 160 32

Year 3: NUMBER - Fractions

Page 101: 1) $\frac{9}{10}$ 2) $\frac{8}{10}$ 3) $\frac{3}{10}$ 4) $\frac{4}{10}$ 5) $\frac{3}{10}$ 6) $\frac{5}{10}$

Page 102: 1) 16 2) 9 3) 5 4) 5 5) 3 6) £54

Page 103: 1) 10 2) 30 3) 9 4) 18 5) 45

Page 104: 1) yes; appropriate explanation 2) no; appropriate explanation 3) appropriate explanation 4) no; appropriate explanation

Page 105: 1) 5 2) 5 3) 3 4) 14 5) 19; appropriate explanation

Page 106: 1) $\frac{2}{3}$ 2) $\frac{3}{4}$ 3) $\frac{4}{5}$ 4) $\frac{5}{7}$ 5) $\frac{7}{8}$

Page 107: 1) $\frac{1}{3}$ 2) $\frac{2}{5}$ 3) $\frac{2}{4}$ 4) $\frac{3}{6}$ 5) $\frac{1}{8}$

Page 108: 1) $\frac{3}{4}$ 2) $\frac{2}{3}$ 3) $\frac{5}{6}$ 4) $\frac{1}{5}$ 5) $\frac{1}{5}$ 6) $\frac{6}{8}$

Page 109: 1) $\frac{3}{4}$ 2) the cat ($\frac{2}{3}$) 3) Isiah; appropriate explanation 4) Humma had the most, Siddiq had the least 5) $\frac{1}{8}$, $\frac{3}{8}$, $\frac{4}{8}$, $\frac{7}{8}$

Page 110: 1) 100 ml 2) £5 3) 5 4) 10 5) 2 kg 6) $\frac{1}{2}$ of £24 circled

Page 111: 1) $\frac{7}{8}$ 2) $\frac{2}{6}$ 3) They have the same amount; appropriate explanation 4) 5 5) no; appropriate explanation 6) £28

Page 112: 1) 5 2) appropriate shading 3) 0.5 4) $\frac{3}{8}$ 5) appropriate explanation 6) £36

MASTERING - Fractions

Page 113: 1) Sarah 15, Kyle 6, Siddiq 3 2 a) and b) appropriate answer given c) $\frac{1}{6}$, $\frac{2}{6}$ (or $\frac{1}{3}$)

Page 114: 3) 9 o'clock 4) $\frac{7}{10}$ = 7 squares, $\frac{4}{10}$ = 4 squares, $\frac{6}{10}$ = 12 squares, $\frac{2}{10}$ = 4 squares

Page 115: 5 a) correct b) not $\frac{1}{3}$ but $\frac{1}{4}$ c) not $\frac{1}{2}$ because they are not an equal size d) not $\frac{1}{5}$ as there is the space in the centre of the shape to take into account

Page 116: 6 a) pipe 2 b) because there is $\frac{2}{3}$ of it hidden but only $\frac{1}{2}$ of pipe 1 and $\frac{2}{3}$ is greater than $\frac{1}{2}$ 7) 10 8) $\frac{1}{10}$, $\frac{3}{10}$, $\frac{2}{5}$

Page 117: 9 a) red = $\frac{12}{24} = \frac{1}{2}$ blue = $\frac{4}{24} = \frac{1}{6}$ green = $\frac{8}{24} = \frac{1}{3}$ b) yes c) red = $\frac{8}{20} = \frac{2}{5}$ blue = $\frac{4}{20} = \frac{1}{5}$ green = $\frac{8}{20} = \frac{2}{5}$

Page 118: 10) any combination where the numerators total 9 11) $\frac{75}{100}$ 12) no because the shapes are not of an equal size, if they were then 2 boxes would have been shaded to show $\frac{1}{3}$

Page 119: 13 a) 90 g b) 160 ml c) 350 m 14) 18 15 a) $\frac{1}{3}$ 15 b) the fraction that is shown as $\frac{1}{3}$ is half the length of the remainder and as this is divided in half the shaded part must also equal $\frac{1}{3}$.

Page 120: 16) appropriate squares shaded 17 a) $\frac{9}{10}$ b) $\frac{7}{10}$ c) $\frac{2}{10}$ d) any digit combinations which total 10

Page 121: 18 a) Monday 3, Tuesday 4, Wednesday 5, Thursday 3 b) 6 c) $\frac{1}{4}$

Page 122: 19 a) $\frac{1}{10}$ b) shapes divided into 8 and 6 equal parts c) $\frac{3}{10} < \frac{3}{6} > \frac{1}{8}$

Page 123: 20 a) $\frac{6}{10}$ b) $\frac{12}{10}$ 21 a) 12 b) 6 c) $\frac{1}{2}$ d) $\frac{1}{5}$ 22) $\frac{1}{10}, \frac{1}{7}, \frac{1}{5}, \frac{1}{4}, \frac{1}{3}, \frac{1}{2}$

Page 124: 23) e.g. $\frac{5}{10} - \frac{3}{10} = \frac{2}{10}$ 24) the answer is the same, any equivalent fractions that are given 25 a) 5 b)10,15

Page 125: 26 a) $\frac{2}{5}$ this is not equal to $\frac{1}{2}$ b) $\frac{3}{9}$ this is not equal to $\frac{1}{4}$ 27) 25 28) a) 8 squares shaded b)15 squares shaded 29) 8

Page 126: 30) 27, 9, 16, 7, 22 31) a) circle 3 stars b) shade 9 of the remaining stars c) 3 stars remaining

Page 127: 32 a) $\frac{1}{4}$ b) $\frac{3}{4}$ 33) 5 34) 9

Page 128: 35) none

Year 3: MEASUREMENT

Page 129: 1) Isiah's 2) Siddiq's 3) Jessica 4) kitchen 5) Kyle 6) May, July

Page 130: 1) Sarah's 2) red 3) glass-top 4) dad 5) the tiger 6) Mrs Whitford's is the heaviest, Mrs Jackson's is the lightest

Page 131: 1) Humma's 2) Humma 3) green jug 4) Fish Tank B 5) Isiah 6) Siddiq's has the most, Sarah's has the least

Page 132: 1) brown 2) mum 3) Fence B 4) Hoxley 5) park 212 m, 9 m 6) Jessica, Humma

Page 133: 1) 900 cm 2) 141 cm 3) 27 cm 4) 1 m 19 cm 5) 47 m 6) 22 cm

Page 134: 1) 100 g 2) 35 g 3) 75 g 4) 215 g 5) 15 kg 6) 2 loaves, 300 g left

Page 135: 1) 30 ml 2) 230 ml 3) 100 ml 4) $4\frac{1}{2}$ litres 5) 50 litres 6) 23 litres

Page 136: 1) 84 cm 2) 300 ml 3) 162 cm 4) 102 km 5) 120 cm 6) $9\frac{1}{2}$ km

Page 137: 1) 50 m 2) 111 cm 3) 625 ml 4) 17 cm 5) 135 cm 6) 1750 g

Page 138: 1) £1.90 2) £4.35 3) 60p 4) 35p 5) £1.80 6) key ring, ball, pen set

Page 139: 1) 60p 2) 35p 3) £8 4) £2 5) £1.10 6) no; appropriate explanation

Page 140: 1) £1 2) £1.70 3) £2.50 4) 55p 5) 90p 6) £2.30 7) £4

Page 141: 1) 255p 2) £3.70 3) 30p 4) £1.90 5) Big Wheel and Twister; Dodgems, Twister and Ghost Train; Space Stroller and Dodgems

Page 142: 1) 120 seconds 2) Isiah, 10 seconds 3) yes; appropriate explanation 4) Isiah's; appropriate explanation 5) 240 minutes

Page 143: 1) appropriate explanation 2) no; appropriate explanation 3) no; appropriate explanation 4) no; appropriate explanation

Page 144: 1) 60 seconds 2) 15 seconds 3) 120 seconds 4) $1\frac{1}{2}$ minutes 5) 300 seconds 6) 240 seconds

Page 145: 1) 31 days 2) July 3) February, November, May 4) 61 days 5) 60 days 6) 122 days

Page 146: 1) 365 days 2) 4 years 3) 366 days 4) 304 days 5) 93 days 6) 323 days

Page 147: 1) 120 seconds 2) 2 pm 3) 6 minutes 4) 4.10 pm 5) 12 noon 6) 1 hour 50 minutes

Page 148: 1) 60 minutes 2) 25 minutes 3) 50 minutes 4) 25 minutes 5) 7 hours 6) 15 minutes

Page 149: 1) 10:50 am 2) 9 am 3) 8:45 pm 4) 12:40 pm 5) 12:15 pm 6) 4:10 pm

Page 150: 1) Isiah's, 5 minutes 2) Sarah 3) Isiah 4) train to the countryside 5) English 6) running

MASTERING - Measurement

Page 151: 1 a) 380 cm b) 120 cm c) 3 2 a) $\frac{1}{4}$ m, 260 mm, 27 cm b) 2 cm, 20 mm

Page 152 3) 10 cm, 5 cm 4) >, =, > 5 a) 90 g b) 180 g

Page 153 6) no; ½ an hour is 30 mins 7 a) $\frac{3}{4}$ b) $\frac{5}{8}$ c) 500 ml 8) no; because 7 x 150 ml = 1,050 ml which is greater than 1 litre

Page 154 9) Isiah £25, Kyle £40 10 a) 4 cm b) 6 cm by 2 cm or 5 cm by 3 cm or 7cm by 1cm c) 6 cm by 2 cm

Page 155 11 a) 50p + 20p, 50p + 10p, 50p + 5p, 20p + 10p, 20p + 5p, 50p + 50p, 20p + 20p, 10p + 5p, 10p + 10p, 5p + 5p b) 55p c) book - £6.30, magazine - £3.15

Page 156 12) pencil - £1.20, pen - £2.40 13 a) 1.30 b) 8.45 c) 11.30 d) 4.45 (accept other appropriate answers)

Page 157 14 a) 4 b) train at 12:00 - 12:13, 12:25, 12:37, 12:50, 1:00 train at 1:30 - 1:43, 1:55, 2:07, 2:20, 2:30, train at 2:47 - 3:00, 3:12, 3:24, 3:37, 3:47 c) 12:25 d) 35 mins

Page 158 15 a) 176 m b) 880 m c) 120 m 16) 50 + 20 = 70p, 50 + 10 = 60p, 50 + 5 = 55p, 50 + 2 = 52p, 50 + 1 = 51p, 20 + 10 = 30p, 20 + 5 = 25p, 20 + 2 = 22p, 20 + 1 = 21p, 10 + 5 = 15p, 10 + 2 = 12p, 10 + 1 = 11p, 5+2 = 7p, 5 + 1 = 6p, 1+2 = 3p

Page 159 17 a) 40 minutes b) 10 minutes c) 14 minutes d) 7:15

Page 160 18 a) 10 minutes 20 seconds b) 10 minutes 50 seconds c) 1st Kyle, 2nd Humma, 3rd Isiah, 4th Sarah, 5th Siddiq

Page 161 19 a) butter 150 g, sugar 180 g, 4 eggs 200 g, 1 egg 50 g

Page 162 19 b) flour 540 g, butter 300 g, sugar 360 g, eggs 8 c) 20 p d) yes; £2.40 + £1.60 + 70 p = £4.70

Year 3: GEOMETRY - Properties of shapes

Page 163: 1) 4 2) no; appropriate explanation 3) 12 4) appropriate descriptions 5) 14 6) appropriate objects named

Page 164: 1) straight 2) sphere 3) 12 4) appropriate drawing 5) square, rectangle 6) cube, cuboid

Page 165: 1) no; appropriate explanation 2) rectangle 3) circle 4) 12 5) appropriate explanation 6) appropriate explanation

Page 166: 1) square, rectangle 2) right angle 3) 90° 4) 2 5) no; appropriate explanation 6) 20

Page 167: 1) straight 2) parallel 3) vertical; appropriate explanation 4) 2 5) appropriate explanation 6) appropriate explanation

MASTERING - Geometry

Page 168: 1 a) possible b) not possible c) possible d) not possible

Page 169: 2) quadrilateral with at least one line of symmetry - rectangle, square, with no lines of symmetry - parallelogram, not a quadrilateral with at least one line of symmetry - isoceles triangle, with no lines of symmetry - pentagon, octagon, hexagon

Page 170: 3 a) 64 cm b) rectangle c)16 cm d) 2 x 16 cm and 2 x 13 cm 4) south

Page 171: 5 a) north b) 3 6) no; because an irregular hexagon can have 6 unequal sides 7) lines of symmetry drawn correctly, (there are no lines of symmetry on G)

Page 172: 8) at least one right angle - A, C, D, F, G at least one pair of parallel sides A, D, E, F, G regular shape E

Page 173: 9 a) larger 1 and 3 smaller 2 and 4 b) 4, 2, 3, 1

Page 174: 10 a) 12 cm and 6 cm b) any appropriate answer.

Page 175: 11 a) and b) appropriately shown

Page 176: 12) triangle appropriately drawn 13) cuboid - 2 square faces and 4 rectangular faces, square based pyramid - 1 square face and 4 triangular faces, cylinder - 2 circular faces, rectangular face, triangular prism - 3 rectangular faces and 2 triangular faces, cube - 6 square faces

Page 177: 14) all shapes except shape 3 are octagons, all shapes have at least one right angle except shape 4 15) appropriately drawn with 6 sides

Page 178: 16 a) 2 b) 3, 5 c) 1, 4

Year 3: STATISTICS

Page 179: 1) 2 pm 2) 3 3) lunch 4) 25 5) no; appropriate explanation

Page 180: 1) 8 2) 0 3) 5 4) 7 5) 30

Page 181: 1) 3 2) 9 3) 4 4) yes; appropriate explanation 5) 24

Page 182: 1) netball 2) 6 3) 7 4) 15 5) 10 6) 3

Page 183: 1) Siddiq 2) 5 3) 3 4) 5 5) 28 6) 22

Page 184: 1) Jessica 2) 8 3) 13 4) 4 5) appropriate explanation 6) 48

Page 185: 1) 6 2) 2 3) 10 4) 0 5) 4 6) appropriate explanation

Page 186: 1) 12 2) Monday 3) 4 4) 12 5) 44

Page 187: 1) black 2) 6 3) 9 4) 4 5) yes; appropriate explanation

Page 188: 1) 12 2) Wednesday 3) 41 4) 11 5) Friday 6) appropriate question

Page 189: 1) 8 2) 5 3) appropriate explanation 4) 22 5) yes; appropriate explanation

Page 190: 1) Jessica 2) Sarah 3) 10 4) 7 5) 1 6) 68

MASTERING - Statistics

Page 191: 1) 32, bar chart completed appropriately. 2) no; 10 children chose mint and 18 chose strawberry, twice 10 is 20.

Page 192: 3) yes; $\frac{1}{5}$ of 120 = 24 4) vanilla and strawberry 5) chocolate 6) 4 7) any appropriate response

Page 193: 8) use the cone to represent 2 or 4 ice creams 9) appropriate answer 10) appropriate reason given

Page 194: 11) 35 12) slide = 17, climbing frame = 28, monkey bars = 14, swings =11, balance bar = 35

Page 195: 13) bar chart completed appropriately 14) the bar chart could not be completed as accurately if it went up in 4s

Page 196: 15) £84 16) a) £16 b) accurately drawn

Page 197: 17) week 5 18) week 3 19) week 1 - $2\frac{1}{2}$ symbols, week 2 - 3 symbols, week 3 - 2 symbols, week 4 - $3\frac{1}{2}$ symbols, week 5 - $1\frac{1}{2}$ symbols, week 6 - 4 symbols, week 7 - $4\frac{1}{2}$ symbols, week 8 - 4 symbols 20) appropriate scale chosen

Page 198: 21) silver 22, black 12, red 8, blue 6, other 10; bar chart completed accurately

Page 199: 22) go up in 2s each time 23) this is for cars that are a different colour to silver, black, red and blue.

Year 3: INVESTIGATION - Trip to Oakham Hall

Page 200 1) £12 2) £3

Page 201 3) 20 minutes 4) 10.45 5) 10.30 6) 15 minutes

Page 202 7) visitor parking to the entrance, funfair to house, house to café all run horizontally 8) visitor parking to entrance is parallel to both funfair to house and house to café. 9) south 10) clockwise

Page 203 11) Sarah's numbers are 65, 70, 75, 80, 85 12) Kyle's numbers are 40, 48, 56, 64, 72 13) Humma's numbers are 20, 21, 22, 23, 24

Page 204 14) Sarah's tickets total is 375, Kyle's ticket total is 280 15) 8, 20, 12, 4, 16

Page 205 16) Kyle throws 4 x 12 and 1 x 8 17) 8 + 8 + 4, 16 + 4, 8 + 4 + 4 + 4, 12 + 8, 4 + 4 + 4 + 4, 12 + 4 + 4

Page 206 18) cylinder 19) 6 20) 9 21) $\frac{2}{3}$ 22) $\frac{1}{2}$

Page 207 23) Sarah scored 4, Kyle scored 3, Humma scored 5 24) Sarah scored 5, Kyle scored 5, Humma scored 4

Page 208 25) 10 minutes 26) 25 minutes 27) 12.40 pm

Page 209 28) quicker to wait for the next train as it arrives at 12:55; if they walk they will arrive at 13:00 29) 8 passengers

Page 210 30) some possible combinations are: 5 5 3 3, 7 7 1 1, 7 5 3 1, 7 3 3 3, 9 5 1 1

Page 211	31) the train with 7 5 3 1 passengers 32) 14
Page 212	33) 12 chairs and 4 tables 34) 3 chairs per table 35) 300 cm
Page 213	36) 10 minutes past 1 37) 13:55 38) 2:00
Page 214	39) S + F + BB, S + M + BB, S + F + P, S + M + P, FF + F + BB, FF + M + BB, FF + F + P, FF + M + P, CN + F + BB, CN + M + BB, CN + F + P, CN + M + P 40) £2.50
Page 215	41) £2.10 42) 40p 43) £2.65 44) sausages, fries and peas
Page 216 & 217	45) week 1 131 meals, week 2 126 meals 46) 89 sausages, 74 fish fingers, 94 chicken nuggets 47) chicken nuggets, sausages, fish fingers 48) 20 more chicken nuggets than fish fingers were sold 49) own choice